Advanced Sciences and Technologies for Security Applications

Indexed by SCOPUS

The series Advanced Sciences and Technologies for Security Applications comprises interdisciplinary research covering the theory, foundations and domain-specific topics pertaining to security. Publications within the series are peer-reviewed monographs and edited works in the areas of:

- biological and chemical threat recognition and detection (e.g., biosensors, aerosols, forensics)
- crisis and disaster management
- terrorism
- cyber security and secure information systems (e.g., encryption, optical and photonic systems)
- traditional and non-traditional security
- energy, food and resource security
- economic security and securitization (including associated infrastructures)
- transnational crime
- human security and health security
- social, political and psychological aspects of security
- recognition and identification (e.g., optical imaging, biometrics, authentication and verification)
- smart surveillance systems
- applications of theoretical frameworks and methodologies (e.g., grounded theory, complexity, network sciences, modelling and simulation)

Together, the high-quality contributions to this series provide a cross-disciplinary overview of forefront research endeavours aiming to make the world a safer place.

The editors encourage prospective authors to correspond with them in advance of submitting a manuscript. Submission of manuscripts should be made to the Editor-in-Chief or one of the Editors.

Kirby L. Wedekind

Constructing Power: A Critical and Constructivist Theory Using the American Case

Springer

Kirby L. Wedekind
Cranfield Forensic Institute
Cranfield University
Cranfield, UK

ISSN 1613-5113 ISSN 2363-9466 (electronic)
Advanced Sciences and Technologies for Security Applications
ISBN 978-3-031-84892-6 ISBN 978-3-031-84893-3 (eBook)
https://doi.org/10.1007/978-3-031-84893-3

This Springer imprint is published by the registered company Springer Nature Switzerland AG
The registered company address is: Gewerbestrasse 11, 6330 Cham, Switzerland

For my family

Preface

This work is a search for answers to questions about how the world works and why it does so. In the beginning, I wanted to understand why the United States, if it is so powerful, continued to fail in what seemed to be rather spectacular ways. The Global War on Terrorism offered the clearest example of this suboptimal performance. My participation in these conflicts offered perspective and a place from which I could begin my search for understanding. I used my experiences to drive my doctoral research, which I have adapted into this book. I found myself in good company upon diving into the many works that have probed this same subject. The foray led me from realism and liberalism—and their many branches—to constructivism and critical theory. I found that constructivism offered a clearer way of understanding the complexity and complications of human interaction, while critical theory provided the sharpest scalpel for purposefully dissecting the implications on the collective society. Constructivist and critical theory provided valid methodological constructs for understanding power and the world in which we live.

In the beginning, I found a call to action from US Senator J. William Fulbright who in 1966 recognized:

> It would be a fine thing indeed if, instead of spending so much of their time playing "war games," political scientists were asking how it came about that we have had for so long to devote so great a part of our resources to war and its prevention, and whether we are condemned by forces beyond our control to continue to do so. The scholar can ask what is wrong with the "other side," but he must not fail to ask as well what is wrong with our side, remembering always that the highest devotion we can give is not to our country as it is but to a concept of what we would like it to be.

I embraced this admonishment and used this work to seek an understanding of the world in which we live, the circumstances that characterize it, and how we view ourselves and each other interacting within a social relation. The work changed me—the looking-glass self manifest. My beliefs, my identity, my interests, and the ways in which I exercise the power in my possession reflect the journey. I return these pages as a testament to my growth as a person and the understanding that I have found about self, other, and the worlds within which we know one another.

Cranfield, UK Kirby L. Wedekind

Acknowledgements

I am indebted to Dr. Anastasia Filippidou for supervising my doctoral thesis that has become this book. Her mentorship and leadership has left an indelible mark on my life and made me into a better version of myself. I am sincerely appreciative of Dr. Matthew Healy, Dr. Gemma Collantes Cellador, and Dr. Anicée Van Engeland for guiding me along this journey and helping make the path clearer. For Prof. Inderjeet Parmar and Prof. Andrew Silke, thank you for reading this work so many times and providing such helpful feedback; I have been amazed by your generosity, passion, and willingness to help. To my parents: thank you for making my education a priority throughout my upbringing. To my wife, Allison—thank you for believing in me and encouraging me throughout this process.

Contents

List of Figures

Chapter 1
A New Conception of Power in Form and Practice

> No one ever spake such warlike words: 'What is good? To be brave is good. It is the good war that halloweth every cause.'
>
> O Zarathustra, our fathers' blood stirred in our veins at such words: it was like the voice of spring to old wine-casks.
>
> When the swords ran among one another like red-spotted serpents, then did our fathers become fond of life; the sun of every peace seemed to them languid and lukewarm, the long peace, however, made them ashamed.
>
> How they sighed, our fathers, when they saw on the wall brightly furbished, dried-up swords! Like those they thirsted for war.
>
> For a sword thirsteth to drink blood, and sparkleth with desire.
>
> Friedrich Nietzsche, *Thus Spake Zarathustra*

This book is about power. Power pervades the world, manifesting within every social relation, shaping the outcome of each engagement, and by extension the world itself. Power transcends multiple fields of academic inquiry, including political science, sociology, and psychology, and serves as one of the critical elements across entire schools of thought despite carrying with it no consistent definition. Kenneth Waltz recognizes in the quote above that power plays a critical role in international relations theory, but its definition and subsequent understandings of power prove contentious [52, p. 46]. Dennis Wrong echoes this sentiment when observing that "Power has always been one of those words that everybody uses without necessarily being able to define satisfactorily. It is treated both as a quality or attribute possessed by individuals, groups, or larger social structures and as an indicator of an active or interactive process or relation between individual or collective actors" [57, p. viii]. Max Weber defined power as "the possibility of imposing one's will upon the behaviour of other persons…[and] the ability of one or more persons to realize their own will in a communal act against the will of others who are participating in the same act" [54, p. 28]. Robert Gilpin views power as "simply the military, economic, and technological capabilities of states" [23, p 13]. Michel Foucault observes that power "is a

K. L. Wedekind, *Constructing Power: A Critical and Constructivist Theory Using the American Case*, Advanced Sciences and Technologies for Security Applications,
https://doi.org/10.1007/978-3-031-84893-3_1

total structure of actions brought to bear upon possible actions; it incites, it induces, it seduces, it makes easier or more difficult; in the extreme it constrains or forbids absolutely; it is nevertheless always a way of acting upon an acting subject or acting subjects by virtue of their acting or being capable of action. A set of actions upon other actions" [20, p. 789]. Hannah Arendt argues for the necessity of establishing the differences between terms that shape conceptions of power: "It is, I think, a rather sad reflection on the present state of political science that our terminology does not distinguish among such key words as 'power,' 'strength,' 'force,' 'authority,' and, finally, 'violence'—all of which refer to distinct, different phenomena and would hardly exist unless they did" [1, p. 142]. Arendt cites Alexander Passerin d'Entreves who posits that: "Might, power, authority: these are all words to whose exact implications no great weight is attached in current speech; even the greatest thinkers sometimes use them at random. Yet it is fair to presume that they refer to different properties, and their meaning should therefore be carefully assessed and examined...The correct use of these words is a question not only of logical grammar, but of historical perspective" (1962, p. 7 in [1, p. 136]). The magnitude of the concept of power offered in these definitions demands a more comprehensive understanding of power, its elements, and its exercise in both theory and practice. Despite its ubiquity and significance, explanations of the concept of power remain incomplete and therefore misunderstood, which creates interdisciplinary confusion between fields when seeking to examine and understand power. John K. Galbraith observes that "in modern industrial society both the instruments for subordinating some people to the will of others and the sources of this ability are subject to rapid change. Much of what is believed about the exercise of power, deriving as it does from what was true in the past, is obsolete or obsolescent in the present [22, p. 3]. While some truth exists within previous definitions or observations, a power model that presents a causal process and explains how power functions does not exist. This creates a problem as different fields and even schools within fields take power to mean something different, the lack of a universally applicable definition of what power is and how it functions creates competing versions of a singular phenomenon resulting in interdisciplinary confusion. The purpose of this work is to define power, its elements, and the process through which it functions in a way that applies across the different fields of academic inquiry and their respective schools of thought. This work seeks to provide a clearer understanding of power than presently exists and in doing so facilitate an increased understanding of self and other while engaged in each power relation.

The success or failure of many endeavours depends upon the exercise of power. International security relies upon power—exercised as both projection and resistance—to achieve or sustain a semblance of order and stability that facilitates security. At present, there does not exist a framework, model, or heuristic for understanding the role that power plays in the relationship between identity, interest, and the exercise of power. This book seeks to rectify this omission by providing such an examination and corresponding framework for understanding power and how it works, which can inform the interpretation of events transpiring within the field of international security by both participants and observers. The purpose of the framework is to enable analysis that accounts for the causal relation between identity, interest, and the exercise of power—ultimately offering greater clarity as to why things happen as they do. The framework that this book offers applies to all types of relationships; while its case studies focus on international relations, the theory accounts for how power transpires between individuals, collectives, and individual-collective relations.

Power operates within every social relation regardless of the environment in which the engagement occurs or the disposition of the agents towards one another; that is, power shapes each interaction between individuals and also shapes the social and physical environment within which each interaction between individuals occurs. Power between individuals reveals agential power while the social and physical environmental within which agents interact constitute structural power. The core of this perspective on power is "As Marx succinctly put it, 'Men make their own history but they do not make it just as they please; they do not make it under circumstances chosen by themselves, but under circumstances directly encountered, given and transmitted from the past" [32, p. 26]. Both agential and structural power operate within every social relation. The intersubjective social-power relation transpires between agents and occurs within context specific to the interaction that includes the past and present social and physical conditions within which the interaction takes place. Agents interacting is significant just as is the set of circumstances that characterize the environment within which any given agents interact.

Power drives every relation along the same pathway, but each iteration produces a unique disposition that reflects the understanding of self and other held by the agents' party to the engagement and their corresponding interests comprising those ideals or things that justify the exercise of power. While the sources, instruments, and forms of power prove finite in theory and limited, the ways in which an agent like the state may combine them seem nearly infinite and reflect the uniqueness of the agents, their engagement, and the circumstances characterizing their interaction. This work offers a heuristic framework that can account for the multitude and diversity of factors influencing the way in which each engagement transpires and its consequent outcome.

1.1 What is Power?

Power is the ability to satisfy oneself. Power shapes every social relation that one enters from birth until death. Each interaction between individuals constitutes a social relation wherein power serves as the operative and determinative gravity; that is, power drives social relations, which necessarily transpires between two actors situated within a particular context of both past and present. Interaction between actors initiates the identity formation process, which creates an identity through role-taking and alter-casting and corresponding interest for each actor. Interest comprises the set of circumstances that an actor desires to create or sustain, the success of which depends upon access to and the exercise of power. The identity and interest formation process transpire within the actor with both endogenous and exogenous influences.

Both aspects combine to constitute the intersubjective social-power relation. Power commands both the formative process within one actor that produces identity and interest as well as the intersubjective social-power relation between two actors; these processes recur concurrently and simultaneously, which makes identity and interest independently fluid and relative to the other. Within every intersubjective social-power relation, each agent draws upon sources of power, adopts an instrument of power, and projects a form of power upon the other agent party to the relation to achieve its respective interest; each projection of power meets with resistance from the other such that the intersubjective social-power relation exists in a state of flux characterized by aggression and resistance. This work defines power as the ability to satisfy oneself.

This book applies its power theory to the state and specifically the United States. National interest derives from national identity, which is the construct that describes the collective perspective, disposition, and belief of the citizenry of a state. National identity shapes state behaviour within the international order through its influence on how the state defines its national interest. Understanding a state's national identity can provide insight into why that state behaves as it does; that is, why the state defines its goals as it does, designs its policies, and how it implements its policies in practice. Where national identity influences national interest and consequently state policy and behaviour, national power determines the means available to the state to pursue its goals, implement its strategy, and achieve its objectives that collectively form its national interest. National interest guides state behaviour, while national power determines the mechanisms through which that behaviour manifests.

1.2 How Does Power Operate?

Every social agent exists in a state of perpetual conflict from its beginning to its end. Interactions between individual agents begin upon birth and continue until death. These relations constitute the intersubjective social power relation, which begin at birth regardless of whether humanity exists in either the state of nature or social contract; further, the establishment of a social contract does not emancipate the individual from the state of nature, but rather adds a structural dimension through which power operates within a society. Within the social contract or beyond it, the sources, instruments, and forms of power remain constant. The social contract only shapes how individuals can access and exercise power, which transpires more apparently through structural forces that manifest within society as organizations and networks than as does the lone individual. The individual never exists in an environment that is not shaped by the other; the individual always exists in an environment shaped by the other. The state of nature described by Rousseau, Hobbes, and many others persists within the social contract and state of reason. Human beings do not emancipate themselves from their nature by implementing social systems of government, but rather burden themselves by doing introducing collectives into society as social individuals that wield structural power against which the individual must compete. This burden manifests as systems of complex organizations against which the individual must work to secure their humanity; that is, the individual exercising agential power and the collective exercising structural power exist in a natural state of conflict within the social contract. While the social contract does allow for the emergence of human rights, it also requires the exercise of power to secure them [15, p. 23, 24, p. 64, 43, pp. 26–27, 44, p. 4, 6, 9, 53, pp. 18–19].

Each interaction constitutes a social relation between subjects wherein power serves as the operative and determinative gravity; that is, power commands every social relation, which necessarily transpires between two and only two actors, whether individuals or groups. Interaction between actors initiates the identity formation process, which creates an identity and corresponding interest for each actor. Interest comprises the set of circumstances that an actor desires to create or sustain, the success of which depends upon access to and the exercise of power. The identity and interest formation process transpire within the actor with both endogenous and exogenous influences while the social relation occurs between actors. Both aspects combine to constitute the intersubjective social-power relation. Power commands both the formative process within one actor that produces identity and interest as well as the intersubjective social-power relation between two actors; these processes recur concurrently and simultaneously, which makes identity and interest independently fluid and relative to the other. As [28] observes: "The new information does not speak for itself but requires interpretation, and interpretation is inevitably influenced by prior beliefs. Of course, if one country attacks another, all observers will take this as evidence of hostility (although the causes of the attack, whether it was provoked or not, and what it tells us about the state's other characteristics can all be disputed)" (p. xlix). Within every intersubjective social-power relation, each agent draws upon sources of power, adopts an instrument of power, and projects a form of power upon the other agent party to the relation in an attempt to achieve its respective interest; each projection of power meets with resistance from the other such that the intersubjective social-power relation exists in a state of flux characterized by aggression and resistance.

Power is the ability to satisfy oneself. Satisfaction arises from either overcoming the other through projection or resisting the attempts of the other to overcome the self; in this way, power manifests as an ability to do and to resist, transpiring between two and only two agents. Three kinds of social relations exist—between two individuals, between two collectives, or between an individual and a collective. The determinative factor in reducing all social relations to two and only two actors emanates from the particular identity that each agent ascribes to the other within the relation; that is, for each agent, the identity of the other is a unique manifestation of the looking-glass self [55], p. 404]. Within power relations, exogenous identity supersedes the endogenous. This drives conflict between agents. Endogenous identity determines the ideal set of circumstances for the agent constituting an interest that requires external action through the exercise of power with the other, but once the relation begins the exogenous identity perceived by the other and applied to the agent factors into the exercise of power that transpires between the two. An exercise of power may seek to reconcile the difference between endogenous and exogenous identity within the relation, but such victories prove fleeting, if ever won at all.

Engagements between agents can happen concurrently, which gives the impression of multiplicity and in such cases these relations require reduction to the agent that possesses a unique identity, interest, and ability to exercise power. An individual agent engaged with a collective does maintain multiple social-power relations simultaneously, but each exists independently from the others. An individual interacting with three individuals comprising a collective bears four unique social-power relations simultaneously because each of the three individuals possess identity and interest of their own, along with a corresponding ability to exercise power within the interaction; the collective which they form together also possesses its own identity, interest, and power arising from their interactions with each other. The individual set against them therefore maintains four distinct manifestations of their own identity, which appear as reflections of the self as perceived by the three individuals independently and the collective they form; the degree of bimodal symmetry within the collective accounts for how the collective formulates an identity and interest while determining how it might exercise power within its unique relation with the external agent—which necessarily differs from the three individuals comprising the group and their respective relation with the other, who must manage four social power relations at the same time.

Defending the power theory presented in this book requires examining the process underlying the intersubjective social-power relation by tracing how it develops from the initiation of the identity and interest formation process, outlining the correlation between the sources, instruments, and forms of power, and describing the way in which an exercise of power transpires between both individual and collective actors. Theories of power have relied upon analysing power in its pure forms despite recognizing that these rarely, if ever, exist. Power proves complex in its machinations and complicated by the inherent subjectivity of humanity seeking to understand itself. This work embraces the chaos that characterizes power from its formation to its use. The structure of this chapter presents its defence of this power theory linearly but only because it remains linguistically necessary to do so. The diagrams that follow each subsection represent the respective aspect of the intersubjective social-power relation, which combined offers a heuristic for analysing power.

1.3 The Relational Properties of Identity and Agency

A power theory requires agents that may possess and exercise power. The term "agents" accounts for both individuals and collectives, with the former describing individual persons and the latter any organized group or party comprising multiple individuals and, which Danto terms as "social individuals," that possess an identity and corresponding interests that an agent seeks to achieve through the exercise of power [12], p. 313]. Emirbayer argues against the conception of agency as "the self-actional notion of 'human will,' as a property or vital principle that 'breathes life' into passive, inert substances (individuals or groups) that otherwise would remain perpetually at rest" and offers a relational perspective that "sees agency as inseparable from the unfolding dynamics of situations" wherein "actors can enter into relationship with surrounding persons, places, meanings, and events...Agency is always a dialogic process by which actors immersed in the durée of lived experience engage with others in collectively organized action contexts, temporal as well as spatial. Agency is path dependent as well as situationally embedded; it signifies modes of response to problems impinging upon it through sometimes broad expanses of time as well as space." [17, p. 294]. Agency conceived endogenously presumes effect without cause. The notion of human will as prerequisite for agency denies it to all other living and non-living entities. The relational conception of agency for which Emirbayer argues associates social interaction as cause with the creation of agency in the individual as effect, it also allows for arguments that would assert agency can exist in other entities, both living and non-living such as animals and organizations, respectively. Social interaction with the other initiates the identity formation process. Lukes proposes that "the very meaning of 'identity' hovers between...the ideas of what is given and what is created or constructed, between finding one's true self and creating it, between self-discovery and self-invention, between primordial identification and post-modernist self-fashioning. One may identify with one's ascribed identity or one may seek to ignore or reject it" [32, pp. 118–119]. Wendt offers that "to have an identity is simply to have certain ideas about who one is in a given situation, and as such the concept of identity fits squarely into the belief side of the desire plus belief equation. These beliefs in turn help constitute interests" [56, pp. 169–170]. Identity constitutes anew in each social relation for the agent and depends upon the nature of the engagement, which includes the time, place, environment, historical context, and the effect that the agent causes to constitute within the other.

1.4 Individual and Collective Identity Constitution

Within every social relation, power causes the action. In this way, "causal and constitutive effects are different but not mutually exclusive. Water is caused by joining independently existing hydrogen and oxygen atoms; it is constituted by the molecular structure known as H_2O" [56, p. 25]. If the metaphor invoked by Wendt holds true, then interaction between agents initiates the intersubjective social-power relation, that is, two agents, or subjects, exercising power against each other through social engagement. Interaction therefore constitutes the intersubjective social-power relation while power binds it. Power therefore cannot exist a priori before socialization wherein identities and interests form or independently from an intersubjective social-power relation. Constitution of agents precedes interaction, the prerequisite and enabling activity for the exercise of power. This indicates that power cannot create without agents or structure created by agents, but once created it commands.

Every agent, whether individual or collective, obtains an identity and corresponding interest through the socialization process "of learning to conform one's behaviour to societal expectations" [56, p. 170]. The agent arrives in a world that precedes it and shapes its identity exogenously and continuously resulting in an identity and interest that proves fluid and relational, that is, identity does not exist for the agent in the same way within each intersubjective social-power relation—on the contrary, the identity and interests of a single agent manifest differently within each intersubjective social-power relation the agent enters. "This principle of identity-formation is captured by the symbolic interactionist notion of the 'looking-glass self,' which asserts that the self is a reflection of an actor's socialization" [55, p. 404]. Given that different individuals interact across space and time, socialization transpires uniquely based on the set of circumstances—both within the physical environment in which the interaction occurs and also metaphysically within the individuals party to the interaction—which indicates that identity proves manifold in its manifestations; that is, interaction shapes a specific rendering of identity and corresponding set of interests for the individual within each separate engagement with the other. While the individual may believe that their identity remains constant, the way in which their identity resonates within the other constitutes a unique manifestation specific to that particular engagement, which suggests that identity in an individual can form from its innate elements in as many ways as a line can bisect a circle. If this holds for identity, then it would also hold for interest formation given the causal relation between the two. Berger [3] describes identity as relational: "Identity, with its appropriate attachments of psychological reality, is always identity within a specific, socially constructed world," upon which [55] elaborates that "each person has many identities linked to institutional roles, such as brother, son, teacher, and citizen…a state may have multiple identities as sovereign, leader of the free world, [or] imperial power. The commitment to and the salience of particular identities vary, but each identity is an inherently social definition of the actor grounded in the theories which actors collectively hold about themselves and one another and which constitute the structure of the social world [3, p. 111, 55, p. 397].

Exogenous constitution of individuals suggests that the Hobbesian state of nature persists within the state of reason and social contract; that is, the innate dispositions of human beings that exist within the state of nature endure and shape the behaviour of individuals in society. "Selves, not to mention the interests corresponding to them, require mutual trust and reciprocal recognition to come into being, conditions that happen to be absent in the state of nature, where a 'war of each against all' prevails" [17, p. 295, 26, p. 78]. Pizzorno argues that "the fiction of individuals not yet involved in social relations but originally knowing what their interests are…is discarded in favour of a view in which the interaction between persons mutually recognizing their right to exist is the only originally conceivable reality. No pre-established interests are imagined. The individual human agent is constituted as such when he is recognized and named by other human agents" [41, p. 220]. Recognition within the intersubjective social-power relation constitutes the individual agent, which allows for and initiates the identity formation process. This characterizes the process as necessarily relational in the sense that "the very terms or units involved in a transaction derive their meaning, significance, and identity from the (changing) functional roles they play within that transaction. The latter, seen as a dynamic, unfolding process, becomes the primary unit of analysis rather than the constituent elements themselves" [17, p. 287]. Identity and interests form through the "dynamic, unfolding process" that Emirbayer describes wherein power serves as the operative, causal force catalysing it. Identities therefore cannot be "assumed as independent existences present anterior to any relation, but…gain their whole being…first in and with the relations which are predicated of them. Such 'things' are terms of relations, and as such can never be 'given' in isolation but only in ideal community with each other" [9, p. 36]. Emirbayer traces the lineage of this line of thinking to Heraclitus and argues that "relational theorists reject the notion that one can posit discrete, pre-given units such as the individual or society as ultimate starting points of sociological analysis (as in the self-actional perspective). Individual persons, whether strategic or norm following, are inseparable from the transactional contexts within which they are embedded" and invokes Foucault to explain that "soul is not a substance; it is the element in which are articulated the effects of a certain type of power and the reference of a certain type of knowledge" [17, p. 287, 19, p. 29]. Foucault further posits that "relations of power are not in a position of exteriority with respect to other types of relationships (economic processes, knowledge relationships, sexual relations), but are immanent in the latter; they are the immediate effects of the divisions, inequalities, and disequilibriums which occur in the latter, and conversely they are the internal conditions of these differentiations" [21, p. 94]. Power, an immanent property of relationships, causes the strife between agents that Bourdieu describes as "the relations of force that obtain between the social positions which guarantee their occupants a quantum of social force, or of capital, such that they are able to enter into the struggles over the monopoly of power" [6, pp. 220–230]. As an immanent power of relationships, power "emerges out of the very way in which figurations of relationships…are patterned and operate." [17, p. 292]. The relational conception of identity and agency offers an alternative to rational-choice theory, which presumes the alliance of interest and action in an individual, and game theory, which adopts a normative process of punishments and rewards in a closed set of individuals [16, p. 28, 17, pp. 284–285]. Elster [16] argues for the relational approach by stating that "the elementary unit of social life is the individual human action. To explain social institutions and social change is to show how they arise as the result of the action and interaction of individuals" (p. 13). Wendt explains the relational in terms of "reflected appraisals or mirroring" that views "identities and their corresponding interests are learned and then reinforced in response to how actors are treated by significant Others," a process that depends upon the beliefs held by one party concerning how the other understands the self-ascribed identity of the first party [56, p. 327]. Wendt elaborates that "if the Other treats the Self as though she were an enemy, then by the principle of reflected appraisals she is likely to internalize that belief in her own role identity vis-à-vis the Other. Not all Others are equally significant, however, and so power and dependency relations play an important role in the story" [56, p. 327]. Reflected appraisals therefore plays a role in the socialization process with the significance of that role correlated directly or inversely to the value that an agent ascribes to the other upon which the reflection falls. If an agent holds a negative perception of the other, it might produce an inverse correlation wherein the agent adopts an opposite behaviour than what the agent views as reflected in the other [10, p. 906]. Socialization occurs regardless of the perception of the other held by the agent and any subsequent action.

Socialization transpires through social institutions and the corresponding laws, in this context as the result of a normative process, that prescribe behaviour in a given society. Mandelbaum [33] advocates for methodological holism: "a whole is not equal to the sum of its parts, and a society is not equal to the sum of those individual activities which go to form it… it follows that the 'parts' of a society are not individual human beings, but are the specific institutions, and other forms of organization, which characterize that society" (p. 231). Wendt invokes an example that Mandelbaum offers on the process of "cashing a check at a bank. In order to perform this action teller and patron must both understand what a check is and what their roles are, and this shared knowledge must be backed up by the institutional context of a bank and banking system" [56, pp. 171–172]. Shared understanding about intangible aspects of society constitutes an agent that influences the behaviour of its members, which indicates that institutions and organizations can possess agency in the same way as do individuals. Societies therefore constitute "multiple overlapping and intersecting sociospatial networks of power," which includes both individuals and collectives [17, p. 295, 34, p. 1]. Implicit in this definition is the emergence of identity for the group, organization, or collective under consideration within a society. O'Connor expands upon Brodbeck's definition of emergence introduced on page 91 by defining emergence as "a (typically) simple, non-structural natural property that is exemplified by objects or systems that attain the appropriate level and kind of organizational complexity and that exerts a causal influence on the behaviour of its possessor" [37, p. 95]. For individuals, the intersubjective social-power relation creates identity that begets agency. Collectives form as a secondary effect of the intersubjective social-power relation, that is, the intersubjective social-power relation must transpire between individuals prior to individuals organizing into groups. The same othering process based on reflected appraisals and mirroring produces groups in the perception of an agent. When the two individuals engaged in the intersubjective social-power relation encounter a third, the othering process begins again among the three individuals. Each agent redefines the situation with respect to their identity relative to their perception of the identities of the two others and groups like individuals together, which may include the self with one of the two others or the two others together. This process reduces the engagement from three agents to two, represented by one individual and one group comprising two individuals. Shared understanding between the members of the formed group begins to shape the identity of the collective which remains fluid and continues throughout the duration of the organization.

The state exemplifies the collective formed through organization and possessing an identity, which combined afford the state with the capability for intentional action. Acting intentionality presumes the existence of shared desires and beliefs in the state about itself and the world in which it exists [56, p. 172]. Searle [45] defines intentionality as "that property of many mental states and events by which they are directed at or about or of objects and states of affairs in the world" (p. 1). Intentionality in this sense distinguishes between collective identity, which the state possesses as shared understanding about itself distributed within the minds of its constituents, with some approximation of collective consciousness, which would imply the consolidation of understanding in a single vessel. Collective identity can only exist as common understanding held by individuals. Given that an individual identity might contain innumerable variables to which the individual might attach their identity, collective identity forms when its members share at least one of these variables as constitutive of their identity and recognize it within the other members. Recognition of shared identity variables cultivates belief of belonging to the collective and so long as this belief holds true in the individual, they remain part of the group. Collective identity organizes the group, while belief sustains it.

1.5 Meaning Making and Interest Formation

Intersubjective social-power relations between ideas held within individuals and by the collectives in which they exist comprise the meaning-making process which precedes interest formation. Belief depends upon the interpretation of sensory data by individuals and the corresponding establishment of norms within the collective. The meaning-making process "connect[s] the community to the objects represented in discourse" by collectively determining "a term's meaning and thus truth conditions are 'owned' by the community, not by individuals…In many cases we depend on the 'testimony' of others, past and present, for access to the objects about which we speak…if we are unsure about the meaning or appropriateness of a mental state, we may take advantage of the 'division of linguistic labour' in society by deferring to experts to explicate our own beliefs" [4, pp. 182–189, 8, p. 184, 42, pp. 227–229, 56, pp. 175–176]. This position argues that meaning and interest form through exogenous interaction rather than endogenous generation: "Thoughts are constituted at least in part by external context rather than solely in the heads of individuals" [4, p. 194]. The circumstances—including the objective elements of the physical environment and other individuals or groups combined with the subjective elements that each represents to the other through symbols, their intended meaning, and their actual interpretation—within which an interaction occurs "determines what meanings we can properly attribute to an agent, and if that context is cultural…then thought presupposes society" [5, p. 23, 7, p. 16, 8, p. 177, 11, p. 354, 39, p. 226, 40, p. 170]. Meaning-making within an individual and communication between entities therefore originates through "contact with the outside world" and "presuppose[s] the world in the sense that they 'are dependent upon the usages of words in a society and cannot be individuated in a context-independent way" [4, p. 200]. The social relation therefore emanates from cognition, the act of thinking that may or may not constitute consciousness, which serves as a prerequisite for the intersubjective social power relation through which two entities assault and resist one another [40, p. 163, 56, p. 175].

Meaning-making for the individual depends on the context—in both an acute-conversational and macro-societal sense—within which that individual finds themselves briefly and exists for a given period. This principle holds true for the collective: "As with behaviour…truth conditions for identity claims are communal rather than individual. It is the 'generalized other' that decides whether the US is a hegemon, not the US by itself, and in that sense the cultural constitution of identity (or subjectivity) is a form of power" [56, p. 178]. Within the international system, social individuals—both states and non-state actors—find meaning through othering; a state or corporation might declare itself as the most or the best, but the group must reflect this declaration back upon the agent for it to hold true. The reflected appraisal constitutes an exercise of power and bears a causal effect on the identity of the agent, whether state or non-state actor.

While exogenous interaction shapes the identity and interest of agents, the entity that exists prior to interaction must possess certain innate dispositions that correlate to the resulting properties or characteristics that arise upon interaction with the other in the process that produces agency. This existence prior to interaction does not constitute agency preceding interaction, but rather suggests that interaction cultivates innate characteristics that remain dormant, latent, or potential within an entity until transformed via the intersubjective social-power relation into identity and corresponding interest.

> Ego and Alter are not blank slates, and what they bring to their interaction will affect its evolution. They bring…material in the form of bodies and associated needs, and representational in the form of some a priori ideas about who they are…exogenously given, self-organizing facts—personal and corporate identities—that act upon and resist the world. These identities have reproduction requirements or basic needs that actors must satisfy if they are to survive…the needs of people as physical and ontological security, self-esteem, sociation, and transcendence…the needs of states as physical security, autonomy, economic well-being, and collective self-esteem [56, p. 328].

Identity therefore forms through exogenous interaction that catalyses endogenous traits into the self. The exogenous interaction that creates identity requires further explanation, which a subsequent examination of the intersubjective social-power relation provides. Within this relation, the interaction between entities that transforms them into agents, identity forms through role-taking, defined as "choosing from among the available representations of the Self who one will be, and thus what interests one intends to pursue, in an interaction," and alter-casting, which projects "a corresponding counter-role [upon the other] that makes Ego's identity meaningful" [56, p. 329]. Role-taking and alter-casting depend upon the beliefs about self and other held by the parties to the engagement, beliefs also take their moral and ethical shape from norms that arise from recurring engagements with entities that share similar characteristics that allow for grouping or categorizing that causes othering. Beliefs therefore determine actions within interactions between agents that shape their identities into normalized roles, which indicates that belief plays a causal role in identity formation and the consequently associated interests and that because the beliefs that an agent holds may change over time so will the identity and interests of the same agent.

Power, specifically access to its sources, instruments, and forms coupled with the willingness to use it, determines whether the agents within an interaction will accept or reject the roles taken and casted: "for an interaction to succeed, in the sense that actors bring their beliefs enough into line that they can play the same game, each side tries to get the other to see things its way. They do so by rewarding behaviours that support their definition of the situation and punishing those that do not. Power is the basis for such rewards and punishments" [2, 56, p. 331]. Taylor [48] observes the power of recognition or withholding recognition and its effect on shaping identity: "projection of an inferior or demeaning image on another can actually distort and oppress, to the extent that the image is internalized. Not only contemporary feminism but also race relations and discussions of multiculturalism are undergirded by the premise that the withholding of recognition can be a form of oppression" [48, p. 25, 35]. Lukes [32] views identity projection as an exercise of power, which he describes as "domination," wherein insufficient recognition comprises both "non-recognition or mis-recognition of ethnic or cultural or religious or geographical identities, which the members of subordinate and minority groups in a society endorse and to which they cleave. Or else people can be seen, and so see themselves, as irredeemably defined by a fixed and unalterable inferior and dependent status and set of roles from which there is no exit." (pp. 118–119). Sustained non- or mis-recognition dominates the subject by reinforcing norms that deny the endogenous elements that precede identity inherent within the individual and the group to which they belong, which "imprison[s] them in a false, distorted, and reduced mode of being [32, p. 119]. Undesired recognition that forces an identity upon a subject can also dominate the identity of an individual or group, while perhaps a correct recognition of identity, those "disinclined to identify with some group or category ascribed to them…are pressured into conformity, public self-ascription ('coming out of the closet') and solidarity. In this way, the identifiers, can dominate all those whose attitudes to the group or category in question are less committed, or else ambiguous, indifferent, or even hostile" [32, pp. 119–120]. Incorrectly recognizing the identity of the other impacts them whether done intentionally or from ignorance, which suggests that exercises of power can and do transpire without intentionality. Asserting that individuals and groups will accept incorrect identities cast upon them presupposes that these individuals and groups have a latent identity that reflects a sense of true self, which suggests the existence of an objective identity and corresponding interest.

Identity forms through the intersubjective social-power relation that constructs it and because it recurs continuously between individuals engaging with one another, identity can and will change. While many interactions transpire similarly as normalized engagement becomes accepted by both agents, no two episodes occur in the same way. Change over time shapes the agents by reinforcing, sustaining, or challenging norms that reflect the Heraclitan observations that "all things change to fire, and fire exhausted falls back into things" and "just as the river where I step is not the same, and is, so I am as I am not," which recognize flux as the fundamental principle operative within the world [25, pp. 15, 51]. Interactions between agents transpires differently due to the ceaseless evolution of the agents and their environment. Learning, both when it occurs and when it does not, accounts for one cause that compels agents to revise their understanding of self, other, and the environment within which their interaction occurs: "interactionists call this 'taking the perspective' of the Other, and for Game Theorists it is essential to the 'I know that you know that I know…' regress that constitutes common knowledge" [56, p. 333]. This learning process does not take place spontaneously but must arise as the result of the intersubjective social-power relation, which drives progress based on the reconstitution of society and culture periodically. The necessary beginning to both arises from interaction between individuals wherein power causes the outcome that comprises identity and corresponding interest. This process manifests as recurrent interaction within which power shapes the result—between individuals at the micro level and at the macro level no less than society itself and the cultures resident within it. Assuming a sort of inertia that negates the necessity for explanations accounting for social and cultural continuity "obliterates the fact that both have to be recreated anew in each generation, often with great pain and suffering. To maintain and transmit a value system, human beings are punched, bullied, sent to gaol, thrown into concentration camps, cajoled, bribed, made into heroes, encouraged to read newspapers, stood up against a wall and shot, and sometimes even taught sociology" [32, p. 122, 36, p. 486]. Both individual and collective identities continue evolving through interactions that power drives, no two interactions transpire in exactly the same way, but do rely upon the same sources, instruments, and forms of power that account for the manifold manifestations of intersubjective social-power relations and consequent dispositions of the agents party to the interaction.

1.6 The Causal Relation Between Identity and Interest

Identity proves manifold in that each intersubjective social-power relation cultivates a unique representation of identity for those party to the interaction. Each identity then causes the creation of a set of interests specific to the interaction that reflects how the agent cultivates its identity within the engagement: "Identities are the basis of interests. Actors do not have a portfolio of interests that they carry around independent of social context; instead, they define their interests in the process of defining situations" [55, p. 398]. Interest serves as motivation defined as "the degree to which human being[s], as a participant in the ongoing social process in which [they] necessarily find [themselves], define a problematic situation as calling for the performance of a particular act, with more or less anticipated consummations and consequences, and thereby his organism releases the energy appropriate to performing it" [18, p. 15, 55, p. 398]. The action inherent in this conception of identity determining interest comes from the power available to or in the possession of each agent. Power therefore drives praxis that constructs identities, interests, and their collective meanings between agents: "Identities and interests are constituted by collective meanings that are always in process. 'The social process is one of constructing and reconstructing self and social relationships,' [so] changing the practices will change the intersubjective knowledge that constitutes the system" [47, p. 93, 55, p. 407]. Participating in collective meanings through acceptance provides actors with the "relatively stable, role-specific understandings and expectations about self" that comprises their identity within a given interaction [55, p. 397].

The causal relation between identity, interest, and the exercise of power arises from a chain that originates in each intersubjective social power relation between two and only two agents, even as each agent participates in multiple interactions simultaneously. A change in identity results in a change of interest, and a change in interest results in a change of how one exercises power, which for a single agent makes the relation between the identity, interest, and the exercise of power causal. The results of this causal process is contingent upon particular engagements between two agents; the identity and interest formation process depends upon role-taking and alter-casting, which forms and reforms identity and a corresponding interest in new ways, even if only slightly or predictably, through interaction between two and only two agents.

Hobbes alludes to the role of power as the catalyst within the intersubjective social-power relation and its ubiquity by citing competition, diffidence, and glory as the "causes of quarrel" and then associating each with gain, safety, and reputation, respectively [26, p. 77]. The oft-cited war "of every man against every man…consisteth not in battle only, or the act of fighting, but in a tract of time, wherein the will to contend by battle is sufficiently known…so the nature of war consisteth not in actual fighting, but in the known disposition thereto during all the time there is no assurance to the contrary," which accounts for all time and therefore suggests the universal presence—in actual or, more commonly, latent form—of the power present within the intersubjective social-power relation [26, pp. 77–78]. Hobbes concluded that no other activity could occur under such conditions, but every activity necessarily does occur within such conditions and in fact occurs because of these inherent conditions that shape identity and interest while affording the agent with the ways and means to satisfy itself.

Interaction with the other transforms the inherent dispositions existing within an individual into its identity, which includes an understanding of the needs and wants that would satisfy it that account for its interests. To achieve these interests, the individual exercises the power available to it against the other, both individuals and collectives, with which the individual interacts. Identity provides agency to the individual that motivates it to seek its interests as an end. The sources of power serve as the means to achieve the interest as an end while the instruments and forms of power provide the ways in which the individual may act to exercise power, which includes the power to do and to resist. The individual exercising power does so purposefully and produces an effect on the other party, which satisfies the criteria of intentionality and effectiveness required for an act to constitute an exercise of power. The effect on the other party appears in terms of extensiveness, comprehensiveness, and intensiveness, which accounts for the number of entities that comply, the number of their activities subjected, and the degree to which the complying entities will endure their domination. Interactions between agents reflects this process. Power arises from interaction and forms the identities and interests of each entity, which then rely upon power to achieve their interests through subsequent engagements with each other encountered. This intersubjective social-power relation exists between entities possessing agency ceaselessly. The case studies that follow examine the intersubjective social-power relation between the United States, which serves as the constant entity across the four cases, and the Islamic State of Iraq and Syria in Libya in 2016; the Russian Federation following its annexation of Crimea in 2014; the United Nations Framework Convention on Climate Change leading to the Paris Agreement in 2015; and the abandonment of American Exceptionalism in favour of America First as the core tenet of American identity that transpired between 2017 and 2021. Each case demonstrates the intersubjective social-power relation in action and reveals how identity, interest, and the exercise of power remain causally bound.

Collective entities like the state follow this same process when forming a conception of national identity, interest, and power. Values held by the citizenry serve as the innate dispositions that inform national identity, from which a concept of national interest arises. The sources of power available to the state differ from those upon which the individual relies, but only in form. The state maintains natural and social resources where the former comprises geography, natural resources, and population and the latter accounts for the economic, political, military, psychological, and informational means available [27, p. 90]. The natural and social sources of power available to the state limit the options available to it in pursuing its interest through engagement with other states and non-state actors. The state exercises the power available to it through instruments and forms that combine to produce an effect, each instrument is a type of weapon system that delivers a form as a gun fires a round against a target. Usage indicates intentionality and produces an effect, though accuracy and precision determine the degree of effectiveness against the intended target. The state wields its instruments and forms against its intended targets comprising both states and non-state actors through four domains of national power—the diplomatic, informational, military, and economic. Theory recognizes these domains as distinct but in practice these domains never exist in pure form. Each domain depends upon the sources of power as its basis and relies on the other domains for mutual support.

1.7 Measures of Power Performance and Effectiveness

Within each intersubjective social-power relation, iterations of the power process transpire continuously. The power process runs both parallel to and within the lines of communications that exist between the two agents; that is, power resides within communication and independent from it simultaneously. For the social-power relation to coalesce, the two agents must engage is some form of communication that includes traditional forms of dialogue—those heard, seen, touched, and even smelled or tasted—in addition to dialogues felt or understood through the processing of sensory data received via traditional dialogue. The difference between the first and second types of communication reflect measures of performance, which count the number and type of repetitions, and measures of effectiveness, which determine how effective these repetitions have been in producing the desired impact on the other. As every communication between agents constitutes an attempted exercise of power, it is the effectiveness of the exercise that reveals when an exercise of power has occurred. Making such determinations requires reconciling the status quo ante with the consequent status quo produced by the engagement. To constitute power, the agent must achieve their desired—or intentional—effect upon the other.

Both words and deeds comprise the data that reveals whether an exercise of power has transpired. For the state, data appears as policy and practice. It is possible that either could constitute an exercise of power independent from the other so long as the agent limits their intent only to the realm of either policy or practice; this seems more likely to occur as the passage of a particular policy intended as political manoeuvring that might disrupt the policy goals of the other, and by extension their implementation in practice. The arena in which practice precedes policy bears an association with systems change internationally and revolution domestically [23, p. 41]. The use of policy constituting exercise of power independently from practice also depends upon the intended target of the activity. The agent seeking to exercise power by establishing or sustaining policy must demonstrate intentionality in both purpose and prey to satisfy the intentionality requirement for constituting an exercise of power. Grand questions arise from this proposition concerning the individualist and collectivist positions, specifically concerning whether it is reasonable to presume one can draw such data from individuals with sufficient reliability and integrity to allow for analysis. In a somewhat ironic twist, collectives offer a better option for interpreting intentionality due to the absence of a collective consciousness because organizations and groups must present their beliefs through media as a way of creating and sustaining the collective itself. The state appears as the optimal collective for such analysis given its imperative to survive, whether endogenously based on its constitution or exogenously in response to threats to its existence.

Following the determination of intentionality, a corollary judgment regarding the effectiveness of the action transpires to ascertain if an exercise of power has occurred. A reconciliation between the status quo ante and the status quo can reveal what has changed and why, with the latter indicating how the power holding agent wielded the elements of power to achieve its intent. An exercise of power produces effects measured as extensiveness, comprehensiveness, and intensiveness; or rather, the number of power subjects, how much of their activity does the power holder influence, and how far can the power holder push the subject while retaining submission.

Every relation is an intersubjective social-power relation. With each, there appears an infinity of possible data points. This framework provides a way of understanding which of the multitude matter and what matters are the intent and effects of the actions taken within the interaction. Actions in this context include all those things that one has done and that which one has left undone. For the individual, this becomes a complicated matter due to inherent subjectivity, bias, and the associated unreliability quotient; for the collective, it is only complex and therefore possible to analyse policy and practice as representations of intent and effect.

1.8 Structural Power and the US Foreign Policy Establishment

Structural power influences the exercise of power and its consequences in a similar way to agential power. This is because collectives and groups exhibit characteristics of both agents and structures; that is, the state displays features of both agent and structure. This characterization reflects the multiplicity of power as both an agent and organization can possess an identity and interest while exercising power in pursuit of satisfying them both; structural power refers to the environment that shapes how these identities and interests form and limit how the individual and collective might exercise the power in their possession. It is necessary in this section to identify the specific elements in the United States that clearly demonstrate how structural power works. "Explaining US grand strategy requires us to break with the realist abstraction of the state from society, and analyse the social content and social sources of geopolitical strategy," which requires considering the identity of the decision-makers, what their respective and shared interests might be, and how these identities and interests link the decision-makers to "particular social groups and classes" (De Graaff & Van Apeldoorn, [13, p. 30]. Examination of the US foreign policy establishment as a structural influence on both the domestic affairs and international activities of the United States is intended to fulfil this requirement as the agential power of the state and United States has been addressed in the previous sections within this chapter.

Critical political economic theory presents a perspective that "[contrasts] both neorealism and neoclassical realism, in which state power is narrowly conceived as the accumulated material capabilities of the 'state-as-actor'... [by seeking] to examine the social sources of that power, shaping its social purpose" [13, p. 33]. The key element is the relation between structural and agential power.

> We: (i) see agents as operating within pre-existing social structures that form the (unacknowledged) conditions of their actions, and (ii) see these agents as reproducing or transforming these structures over time. In order to link structure to agency we will understand structure as referring to: (i) the social *position* an agent occupies, that is, within a set of social relations, and (ii) the wider context in which she operates and to which her strategic action may be oriented. Furthermore, *how* structure conditions agency is often mediated by the ideas and beliefs that the agent holds. Ideas cannot be reduced to the social position an agent occupies, yet neither can they be understood as existing independently from that position, that is, from the social structure in which any ideational practice is embedded (De Graaff and Van Apeldoorn, [13, p. 33].

This explanation of structural and agential power reflects the principles at work within the identity and interest formation process as causal elements that drive subsequent exercises of power. The agent enters into a pre-existing environment that shapes beliefs about self and other, though the agent is empowered to alter their beliefs and by extension the environment. Identity is constantly constructed through interaction. The type and kind of others with which the agent interacts constitute the structure of the environment; the entirety of others taken as a collective combine to create a tapestry that limits and empowers the agent. It is necessary to consider both the agent and the structure in which the agent operates to account for identity, interest, and power as each of these elements influence the others. Reciprocation characterizes the process in a sort of harmonization where every aspect both gives and takes from the elements comprising the power process.

For the state and the United States, this perspective requires considering both the United States as agent and the structure that shapes this agency. Walt [51] identifies the US foreign policy establishment—defined as "those individuals and organizations that actively engage on a regular basis with issues of international affairs" including both government and non-governmental organizations and individuals—as one of the key structural forces that shapes United States national interest and exercises of power (pp. 97–101). This group consists of the national security and foreign policy bureaucracy, which "includes the president, vice-president, National Security Council, the relevant personnel from the Departments of State, Defense, Energy, and Treasury, the various intelligence services, the relevant congressional committees, and research agencies like the Congressional Budget Office" and the array of national laboratories [51, p. 96]. All together the government components of the US foreign policy establishment spend hundreds of billions of US dollars annually; nearly one million individuals who work in government participate in crafting or implementing the foreign policy of the United States (ibid). The non-governmental groups that contribute to the foreign policy establishment include the 1800 think tanks, many leading academic institutions, and special interest groups advocating for various causes [51, p. 101]. This collective, the US foreign policy establishment, possesses its own identity, interest, and exercises power. Its identity arises from role-taking and alter-casting—an act that transpired in the preceding sentences that noted the various groups that comprise the foreign policy establishment. Linking these groups together by common purpose, or rather interest, demonstrates that the process through which identity, interest, and the exercise of power flows in both directions. Identity produces an interest, but it is also possible for interest to create an identity that binds individuals and groups to a shared purpose or belief. The interest that connects the many groups comprising the foreign policy establishment is two-fold; the first component is advocacy for sustaining the liberal world order, or liberal hegemony, while the second is self-preservation. Walt [51] defines liberal hegemony as:

> [Seeking] to use American power to defend and spread the traditional liberal principles of individual freedom, democratic governance, and a market-based economy. The strategy is one of hegemony because it identifies America as the indispensable nation that is uniquely qualified to spread these political principles to other countries and to bring other states into a web of alliances and institutions designed and led by the United States. Not only do its proponents see the preservation of US primacy and the expansion of a predominantly liberal world order as essential for US security and prosperity, in their eyes, this objective is good for the rest of the world as well (p. 14).

Advocacy for liberal hegemony serves not only the interests of the United States but also those individuals and groups comprising the foreign policy establishment: "Leading members of the foreign policy establishment undoubtedly believed that liberal hegemony was the right strategy for America, but they also understood that it was very good for them. Open-ended efforts to remake the world in America's image gave the foreign policy establishment plenty to do, appealed to its members' self-regard, and maximized their status and political power" [51, p. 15]. This observation echoes [28] who invokes Upton Sinclair's conclusion that "It is difficult to get a man to understand something when his salary depends on his not understanding it" (p. lxxx). The foreign policy establishment does not solve problems because doing so would reduce the power, influence, and prestige of the organizations comprising it as well as reduce the income, relevance, and sense of purpose felt by the individuals who work in these organizations. To stave off these negative consequences of success, the foreign policy establishment inflates threats, exaggerates benefits, and conceals the costs of its activities and those taken in support of the policies for which it advocates [51, p. 138].

An unlikely union between the realism of Walt and Jervis with the critical theory of De Graaff, Van Apeldoorn, Layne, Vucetic, and others emerges from mutual recognition of the foreign policy establishment or elite as one source of the problem plaguing US foreign policy in practice. De Graaff and Van Apeldoorn [13] observe that "In the case of the United States, its liberal capitalist political economy is characterized by a structure at the apex of which we find the oligarchic top segment of an autonomous capitalist class. The US political system is consequently structurally biased towards the interests of the corporate community, secured through a persistently revolving door and through the so-called policy planning process" (p. 116). The state itself becomes merely a product of the individuals comprising the small population that controls or influences policy rather than a reflection of all the individuals and social groups that exist within it. The result is a state that exercises power according to the smaller population of elites. Where the activities of the state serve the rest, it is only a consequence of happenstance alignment of interest between the several groups rather than deliberate consideration of the others. "Within the USA…[foreign] policy formation largely takes place within an elite network of corporate links, as well as ties to corporate-funded and directed think tanks and other policy planning bodies. This power structure research thus suggests that at the apex of US society we find a corporate elite that is closely networked with both the world of think tanks and with the state, and extends into global transnational elite networks as well" [13, p. 339].

The number of think tanks and organizations from which the Clinton, George W. Bush, and Obama administrations pulled individuals to serve in key roles within their respective administrations nearly equals the number of connections to organizations shared between across each, which reveals the structural power of the foreign policy establishment through its organizations and their ability to transcend changes in administration. This ability to sustain power is the result of agential power wielded by these organizations; sustained influence by multiple organizations creates an environment that limits or empowers the other agents acting within in thereby constituting structural power. The culture that forms endogenously within each organization will reflect its priorities and existing membership so that certain individuals possessing identities and interests that mirror those of the organization are more likely to join it. The exogenous consequence is that the influence wielded by the organizations comprising the structural power of the foreign policy establishment "takes place through several channels such as shaping public and elite opinion/discourse and setting the policy agenda through the media and through more specialized reports and advice, as well as through regular personal contacts between state officials and key business associations and other forms of corporate lobbying" [13, p. 33].

The character of the foreign policy establishment in the United States reflects the composition of the various organizations comprising it rather than the United States itself. Domhoff and Zweigenhaft [14] and Vuctetic [50] argue that white, Anglo-Saxon, Protestant individuals hold significant control over these institutions. Parmar [38] notes that "Anglo-Saxonism is a culturally assimilative force that absorbs elite members of minority communities into dominant mindsets and roles, and maintains existing patterns of power…Elites play key roles and derive greatest benefits from their activities within structural contexts not of their immediate making" (pp. 257–258). The identity—race, social class, religious beliefs, sex, and gender—of the individuals comprising the US foreign policy establishment influences how it interprets events that require action. Vuctetic [50] observes that

> Faced with an event, elites may explain it using dominant discourse or provide a frame based on an alternative discourse. Ignoring it or calling it irrelevant is of course possible, but it is unlikely for the types of events described above because the authority of elites partially depends on their ability to provide continuous and consistent identity narratives in all circumstances. For this reason, as major events and acts emerge and reality constraints change, elites must reappraise state or national identity anew, in which case old narratives may prove unreliable, sometimes even compared to those authored by significant Others (p. 14).

This provides an explanation for the ability of the US foreign policy establishment to survive—it is in its own interest to remain consistent if only to shield itself from the criticism associated with swift change in the face of uncertain circumstances.

The consequence of the structural power of the US foreign policy establishment is continuity in terms of identity and interest across Presidential administrations and consistency in how the United States has exercised its national power despite changes within itself and the international order. Layne [30] outlines the principles that form the Open Door Theory that seeks to explain the key drivers behind the foreign policy of the United States: (1) economic expansion, specifically into overseas markets; (2) economic openness, or ensuring that the overseas markets upon which economic expansion depends become and remain open to US capital; (3) political openness, or rather the practice of supporting foreign governments that implement policies aligned to the interests of the liberal international order and the United States; (4) externalization of evil, which is most clearly viewed as role-taking and alter-casting wherein the United States takes the role of good for itself and casts all opposed as evil; and (5) American exceptionalism, the idea that the United States is different from other states, driven by morality rather than self-interest, and thereby warrants a special status within the international order (pp. 29–36; [13, pp. 35–36]). To achieve these five goals, the United States must "create a stable international order that is safe for the economic Open Door" and doing so requires achieving hegemony through military supremacy because the type of power upon which the use of military force depends—condign power through the use of force and violence—is relatively easier for the United States to achieve given its access to the three sources of power and the reciprocal difficulty in achieving economic, diplomatic, or informational supremacy [30, p. 125]. Since the end of the Second World War, "American military power [has been] considered by US policy-makers and many security studies scholars to be the geopolitical trump card—no pun intended—that will ensure continuing American dominance even if China closes the economic and technological gaps separating it from the United States" [31, pp. 94–95]. Despite the principles of the Open Door setting conditions that favour it, the US foreign policy establishment seems to have failed to recognize the inevitable increase in power held by the states that own or operate the foreign markets upon which the Open Door depends. The result is suspicion and fear of American decline relative to the other, reports of which have ruminated for decades though "primacists argue that the extent of China's rise—and hence of America's decline—are, like premature reports of Mark Twain's death, greatly exaggerated. Primacists believe the international system is still unipolar, and that US power will keep it that way for a long time to come" [31, p. 95]. A gain by the other does not constitute a loss for the self, it is necessary to distinguish between relative and absolute when considering the possession of power whether latent or kinetic. A gain by a rival encroaches only on relative power and does not diminish total power. This holds true equally for the United States as agent as it does for the think tanks and policy organizations, both individually as agents and collectively as the structural power constituting the US foreign policy establishment. Considering further that "Decline is Un-American," it remains that for the United States and any others who might become the hegemon "Whatever happens will be for the worse. Therefore, it is in [their] interest that as little should happen as possible" [31, p. 95].

1.9 A New Constructivist Model for Understanding Power

To aid in explaining the definition of power as the ability to satisfy oneself and its supporting theory that views a causal relationship between belief, identity, interest, and the exercise of power, it is necessary to introduce a power model that can reconcile theory with practice as observed in the case studies that follow in subsequent chapters. The power model is a product of cross-referencing the observations of theory and case study both respectively and collectively, and vice versa; that is, reconciling the theory and case studies against each other during their creation protected their integrity against analytical bias and foregone conclusions. This iterative process is important to note because it is the information uncovered through research—a combination of theoretical inquiry and applied practice—that determined the composition of the model. The model focuses on the agential power within the intersubjective social-power relation while attempting to account for the influence of structural power. As the author's own, this model represents an original contribution to knowledge through increased understanding of what power is and how it operates. The following pages contain few references because this explanation of the model and the power theory that it describes is intended as an original contribution to knowledge (Fig. 1.1).

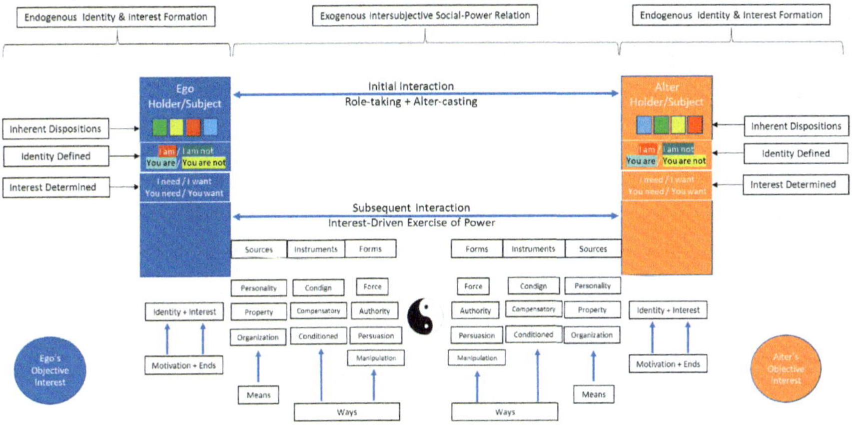

Fig. 1.1 A new constructivist power model (Author's own)

This power model associates the elements of identity, interest, and the exercise of power in a causal flow. The model considers how both agents understand themselves in relation to the other, how this understanding of self and other informs the development of conditions that each believes will serve itself best, and to how the agents might exercise power against each other. Beginning at the top of the model and proceeding to the bottom, the exogenous interaction initiates the sequence as ego and alter first engage in role-taking and alter-casting; this process begins the endogenous identity and interest formation process within which each defines its identity for this interaction relative to the other through a combination of meaning-making derived from this interaction and remnants from previous iterations of this process with others that form what one might perceive as inherent dispositions. Once identity forms, an idea of what one desires takes shape alongside a reflected understanding of what the other desires. The exercise of power then follows, wherein each agent reconciles the desired ends with the means and ways available—the sources, instruments, and forms of power through which the agent may act against the other to achieve the interest derived from identity; the term action in this context includes both the projection of power and its resistance. Power therefore is the ability to satisfy oneself in a given social-power relation by overcoming or resisting the other.

The model outlines the intersubjective social-power relation that serves as a heuristic for understanding more completely the causal effect of power transpiring between individuals and collectives, including both the state and non-state actors. While the individual and the collective can possess power, its existence depends on the same identity-interest formation process and the same sources, instruments, and forms. Power does not belong only to the state, though the state perhaps possesses the greatest access to the sources of power; all other types of collectives, or non-state actors in the international community, like corporations, intergovernmental organizations, and nongovernmental organizations acquire and wield power through this same process. This work holds that power operates this same way between individuals but leaves exploration of this idea for future research that would apply this same power theory and model. The focus here remains the power relations between collectives, specifically the state.

Each initial interaction begins the identity formation process in which each agent cultivates a unique representation of its own identity relative to the other drawn from the set of inherent dispositions or traits that pre-exist within itself described as role-taking. By taking a role, each agent projects its expectations of self and other upon the other through alter-casting. Interaction between individuals initiates role-taking and alter-casting that produce the identities of each agent particular to their engagement; while each agent has many innate and learned traits that comprise the full set of potential identity elements, the way in which identity forms depends upon the relation between the parties to a given interaction [56, p. 329]. The identity formation process creates an understanding in each individual that defines what self is and is not, as well as what the other is and is not. It seems unlikely that any two individuals have ever shared an objectively true understanding of self and other given the inability to form shared consciousness, which distinguishes an individual from a group of individuals regardless of like-mindedness. Individual persons possess a certain degree of determination in pursuing their interest that reveals itself explicitly and implicitly through actions and statements; for a group of individuals, observing and measuring determination requires examining actions and statements performed by individuals acting as agents for the group rather than their individual selves. The degree of determination present within a group reflects the intensity, comprehensiveness, and extensiveness of belief in the mission of the group and willingness to perform actions in pursuit of this mission [57, p. 18]. Individuals within a collective will possess unique understandings about the group and its purposes based on the way in which their identities have formed within the context of the group through role-taking and alter-casting that includes labelling members of the group prior to the initial interaction, which thus may transpire between individuals, between collectives, or between individuals and collectives. Individuals serve as agents of collectives and represent the whole within interactions involving collectives.

Identity formation begets interest formation. Based on the understanding of self and other, each agent formulates a set of goals that together comprise its singular interest, understood as the conditions that satisfy the desires of the agent in terms of needs and wants. The agent believes that it ought to achieve its interest, however defined, to support the role-taking and alter-casting episode preceding it. Identity causes interest, and interest depends upon identity to conceive the set of circumstances that the agent should seek to bring about within the interaction with this and every other. Interest, like identity, depends upon the particular interaction within which it forms. Possessing an understanding of self and other in terms of needs and wants, and informed by an idea of its desires, the interaction moves to the third phase in which the agents employ the sources, instruments, and forms of power against one another to serve or satisfy its interest while confirming or sustaining its identity. In this way, identity motivates the agent to act while interest guides those actions by aligning them to an overarching goal that satisfies the agent. The model distinguishes between the subjective conception of interest possessed by the agent and the objective interest of the agent to illustrate the difference between perception and reality.

In each case study, a distinct manifestation of the identity of the United States forms relative to the other against which it acts that originates in the role-taking and alter-casting by both parties to the engagement. The identity of the United States differs when interacting with the Islamic State of Iraq and Syria in Libya, the Russian Federation, the United Nations Framework Convention on Climate Change, and its own citizenry. This phenomenon does not imply the existence of multiple identities, but rather that the concept of identity, especially for a collective like the state, comprises myriad aspects that coalesce when activated by the act of role-taking and alter-casting; that is, every potential aspect of identity may not appear within the identity of an actor within a given relation with a particular other; that is, the identity of each agent capable of possessing an identity manifests differently within every intersubjective social-power relation, which includes recurring relations with the same other. The time that passes between each engagement changes both parties to the relation such that iterative interactions reveal different aspects of identity resident within the agents. This phenomenon offers one possible explanation for why two agents may achieve different results from each interaction; for collectives like the state, it explains how a collective entity can engage with another collective entity along multiple lines of effort simultaneously and achieve both positive and negative outcomes. Examples of this phenomenon transpiring would include instances wherein two states may appear at odds through the lens of public diplomacy and the associated rhetoric but cooperate on counterterrorism operations, intelligence sharing, or economic exchange. Identity draws on the remnants of previous engagements that comprise what one might perceive as inherent dispositions; the complete set of these dispositions may not manifest within the identity of the agent within each relation. The identity that appears within the intersubjective social-power relation is distilled from the set of possible identities comprising the comprehensive identity of the agent. Identity therefore depends upon the other to form and for the specific characteristics that resonate relative to the other within the relation. The dependence of identity formation on the other causes a corresponding determination of interest for the agent that reflects the way in which its identity manifests. The identity and interest of the United States manifests differently in each engagement with external entities because of the engagement itself, which begins with role-taking and alter-casting, flows to identity and corresponding interest formation, and consequent action that constitutes the exercise of power.

As sources of power, personality, property, and organization provide the agent with the means to exercise power against the other. Each source enables the several instruments and forms of power but bears a primary association with each [22, p. 6]. Personality, broadly defined, incorporates all aspects of the self—both the physical body and all cognition—into a singular construct reflective of Weber's charismatic authority. The difference here arises from personality viewed as a source of power as opposed to a type of authority relationship. Weber presents charismatic authority as a type of legitimate authority based on the personality possessed and wielded by an aspiring power holder [35, p. 206]. The Weberian tripartite authority model fits within legitimate authority as a form of power. Personality as a source of power emanates from physical and mental strength, agility, and endurance that fosters a willingness to act, which therefore bears its primary association with condign power as instrument and force as form, where condign power bestows punishments in the form of physical and psychological violence against the other [22, p. 6]. Personality may also enable conditioned power as instrument and either persuasion or manipulation as form, as the success of these endeavours depends upon self-emanating conviction that resides within personality, which differs from identity. The former describes the capabilities of an individual and the various capacities thereof, while the latter pertains to the characteristics or traits that define personhood for the individual; that is, personality accounts for what an individual can do, and identity defines who the individual is. Identity therefore necessarily precedes personality.

Property, or the possession of property, provides the agent with the means to purchase through exchange or transaction, which therefore creates its primary association with compensatory power and authority as instrument and form of power, respectively [22, p. 6]. Property includes the full measure of resources available to the agent, including owned assets and access to fiscal resources like debt instruments, markets, and arbitrage; property also includes mastery of demonstrated or recognized skills and social status or position. An agent possessing extensive property across diverse assets can leverage these resources to exchange property or access with the other for their compliance or submission. Possessing property therefore serves as a source of power by making an exchange possible between agents where one owns a thing and the other desires it, and links to compensatory power as instrument based on its transactional nature. Property as the possession of materiel also enables condign power and force, while property as possession of social status facilitates conditioned power through persuasion and manipulation.

Organization, as in membership in an organization, serves as a source of power by requiring its members to subordinate their own purposes to the collective purpose of the organization and then wielding this initial wilful submission to sustain its domination, or control, over its members [22, p. 6]. Organization in this sense represents the thing become as opposed to organizing, which describes the process of becoming [46, p. 17]. The act of organizing transpires continuously as it proves necessary to create the organization and then also to sustain it, the organization depends upon its constituents and binds them simultaneously. Increased bimodal symmetry, defined as loyalty of the members to the organization measured in terms of subordinating self-interest to the interest of the organization, directly correlates to the strength of the organization and by extension the power available to its members in wielding it against those outside the organization [22, pp. 56–57]. The organization as a source of power relies upon its ability to shape the beliefs of both its members and others, which connects it to conditioned power as instrument and both persuasion and manipulation as forms. An agent may co-opt the organization by invoking its code or mission as justification for an action separate from that of the individual; the inverse proves equally true if advantageous to the agent within a given engagement if the conditions require it to persuade or manipulate the other. An organization depends on belief in its purpose to compel the submission of its membership and depends upon its members to advocate for its collective interest while engaging with others. Advocacy may utilize any of the three instruments of power and take any form.

An organization necessarily exists as a collective entity. This characteristic distinguishes it from the other two sources and injects complexity into power analysis from a methodological perspective. While personality and property can and do belong to individuals, an organization must comprise multiple agents—even if only two agents, with one of these being the individual agent founding an organization with zero members as a ruse. By creating the empty organization, the agent can justify actions taken by attributing their motivation to the interest of the void organization. In this example, bimodal symmetry proves complete.

The three sources of power serve as the means to exercise power while the instruments and forms provide the ways. The ability to exercise power originates from one of the three sources. The agent chooses the way in which they will exercise power by adopting one of the three instruments as a tool to deliver the volley against the other through one of the four forms of power. In cricket and baseball, a pitcher will vary the speed, spin, and location of the ball to disrupt the batter; the pitcher's arm acts as the instrument while the pitch itself serves as the form. A more complete metaphor would specify the pitcher as ambidextrous and therefore possessing multiple instruments, while the "full monty" would cite the teams as collectives that provide its members with an identity relative to the other from which both pitcher and batter derive organization as a source of power, with personality also in play by dictating the strength possible for each pitch or swing and the likelihood of striking out the batter or getting a hit. Pitcher and batter, already possessing an identity as part of their respective teams, re-enter the identity formation process when their engagement begins, which sees teams as organizations and individuals as pitcher and batter set against one another simultaneously in a contest. Just as bowlers vary each throw by changing the speed, spin, and location of the pitch, the batter expects that each ball thrown will differ from the last. Both agents anticipate the actions of the other and by doing so develop an understanding of how their antagonist will act to achieve its interest of getting an out or a hit, respectively. Effective engagements see variations on pitch and swing based on the circumstances on the field tactically and within the game and season strategically. The interaction between pitcher and batter transpires between them as individuals and their teams as collectives simultaneously. Over time, the individuals will depart the team, but it remains—for a time, at least.

The state and the individuals representing it exercise power in practically the same way. National identity and interest forms through interaction that sees role-taking and alter-casting between two states on the grand or strategic level, meant here to describe the echelon of relations in which states and non-state actors interact as social individuals. Within the interaction, the states party to it develop a unique conception of their identity and interest relative to the other. Subsequent interactions revisit the role-taking and alter-casting stage of engagement that allows for states to modify their conceptions of self and other, which accounts for change in their relations but not necessarily, or even hardly, the structure of the international order itself. Changes in the identities of the states party to the relation compels a consequent change in their interests relative to the other. The change in identity and interest warrant the exercise of power as a way for the state to achieve the set of circumstances that it believes most beneficial to it; that is, the state acts in accordance with the identity and interest that the individuals holding positions of legitimate authority have defined for it during the period in which the power holding individuals remain in such positions. Within a state, it may be the case that the conceptions of identity and interest held by the power holders differ from the subjects; in such cases, the identity and interest formation process still transpires between two collectives as described here and outlined in the model. Ideological fissures within the state happen continuously to sustain or reshape conceptions of national identity and the corresponding interest; these fissures happen within the established governance and prove necessary for the state to endure. Ideological fractures create competing organizations or factions that constitute the other relative to the status quo—perhaps inevitably—ante.

Power transpires between all agents in the same way. Each agent, whether individual or collective, draws on the resources available to it, utilizes an instrument to focus the source, and applies power through the form that the agent believes affords it the greatest opportunity for setting conditions favourable to its interest. Exercises of power incorporate multiple sources, instruments, and forms, and must occur intentionally and prove effective in terms of intensiveness, comprehensiveness, and extensiveness. Each exercise reflects one of the four domains of national power, though a correlation between the exercise of power and the reflected domain does not necessarily exist; that is, an exercise of power drawing on organization and property that utilizes condign power applied through force may reflect either the diplomatic or economic domains in addition to the military, which bears a traditional association with condign power and force. The exercise of power requires and comprises two and only two agents where one seeks to dominate and obtain the submission of the other while the other resists to maintain the aspect of itself according to its own dictates. The reciprocal processes of attempted domination and resistance depend upon the same sources, instruments, and forms; the aggressor must overcome the resistance to succeed, while the resistance must sustain its independence from influence projected by the aggressor. Changes in the behaviour or beliefs of the resisting agent that otherwise would not have occurred indicate a successful exercise of power for the aggressor. The intentionality quotient for an exercise of power derives from generic and not specific intent; that is, the aggressor need only to achieve some change in the behaviour of the resistance and not the specific change that it initially set upon. Subsequent exercises of power intentionally undertaken that produce a tangible effect of behaviour or belief may achieve the original intent or may create conditions that compel the aggressor to modify its tactics while maintaining the strategy.

The Taoist yin yang symbol between the two agents in the power model represents the engagement itself as a type of harmony [49, p. 125]. Each agent depends upon the other for the engagement to exist and exercises its power against the other. While one may come to dominate the other, the engagement constitutes the intersubjective social-power relation that transpires between all agents, a single entity will enter many such interactions and in complex situations may engage with many others simultaneously. These interactions prove reducible to the self and other, or ego and alter, or aspiring power holder and reluctant power subject; that is, the agent does engage with multiple others simultaneously but does not do so concurrently. The common denominator to every interaction remains the agent and the other set against it, even when the agent sets out against a collective or vice versa. The malleability of individual as representative of an organization or collective and the collective itself affirms the significance of organization as a source of power. The political official, complete with identity and corresponding interest, exemplifies the ability of the individual to dissolve into the state and substitute its power for their own. The agent uses the state as a vessel to justify the pursuit of their own interest limited only by their respect for the law and social norms. Power exists in a perpetual state of flux as a reconciliation both within the agent as indicated by the identity-interest formation process and between the agent and other through the intersubjective social-power relation.

1.10 Falsifiability of the Power Model

The power model reduces the many components of power to its elements and attempts to identify the causal linkages between them. The power model cannot account for all the factors that drive one actor or another to behave as they do; to do so would require perfect knowledge of each agent and the structural forces that influences their respective decision-making processes. The sum of these forces—agential and structural, endogenous and exogenous, known and unknown—remains unknowable to those participating in the intersubjective social-power relation and those observing it. Jervis [28] recognizes that "theory and explanation need not fill in all the links between cause and effect. Indeed, this is impossible. One can always ask for the links between the links" (p. 14). It is these "links between the links" that elude both the agents in their understanding of self and the observer or researcher seeking to understand why and how power operates. Perhaps the simplest explanation is that it is impossible to know everything about why individuals behave as they do, how the structures that influence behaviour came into existence and what sustains them, and whether the theory inherent within every fact resonates in the myriad actions that comprise what one might describe as reality. It seems likely that researchers will continue to reveal what remains hidden between the links thus falsifying what led them to inquire.

It might be possible to falsify the model by applying it to cases wherein the power holder cares about the engagement but the power subject does not. Jervis [28] observes "When the situation totally determines response, we cannot attribute any characteristics to the person who responds. That a weak actor, either individual or national, gives in to a strong adversary when the issue is minor to him but not to his adversary tells us almost nothing about his goals, beliefs, and willingness to run risks" (p. 35). If asymmetry in the power relation is such that the power holder cares and the power subject does not then analysis of the engagement cannot bear meaningful insight into either the agents or the structural forces influencing their encounter.

It might also be possible to falsify this power model by attributing causality to structural power only rather than a combination of both agential and structural power. The counterpoint would then ask from where the structural power came, which would lead back to agential interaction. At this point, it reminds of the alleged dialogue between Bertrand Russell and a refutation provided to him following a lecture on astronomy "who claimed that the earth was situated not in space, but on the back of a large turtle. When Russell asked what the turtle was standing on, the woman replied that it was 'turtles all the way down'" [29]. The challenge of infinite regression appears in the social sciences just as it does in the physical sciences. The search for causality in the world makes it so. The woman's belief in turtles all the way down is simply a metaphor for those things that one cannot yet explain or understand. If seeking to falsify this work, one could target any aspect of the power model and ask from where this or that comes. Answers could range from the inherent dispositions driven by genetic heritage and deoxyribonucleic acid to the environment within which one was raised, nature and nurture both influence identity. The choice of mate, career path, home, parenting style, and many other factors begin creating opportunities and placing limitations on progeny before they exist and continue throughout their lives. Individuals are borne into a world that proceeds them, but upon their arrival each has an opportunity to formulate an identity, set of interests, and exercise the power available to them as they can within the world such as it is. The structure that an individual encounters finds its origins in the accumulation of interactions over innumerable generations, though an explanation for its present manifestation in specific places and within certain timeframes might look relatively nearer to the present for the causes of the structure. If nothing else, the power model offers a sketch of the elements of agential power while recognizing that even elements consist of components further and further reducible.

1.11 Future Research and Limits of the Power Model

This work designs a power theory and applies it to the United States and recent episodes. Future research using the model could examine non-state actors or structural forces. Intergovernmental organizations such as NATO and the UNFCCC receive attention here but the emphasis on the United States rather than either intergovernmental group provides an easy segue. Intergovernmental organizations consisting of states from the Global South like the African Union, Gulf Cooperation Council, and Association of Southeast Asian Nations would test the model to see if its components and organization account for how power operates beyond the West. Analysing cases emanating from the Non-Aligned Movement using the power model is a tantalizing prospect.

Using the power model to focus upon structural forces like race, class, and gender would shift the primary focus from agential to structural power. The model could be expanded to include the influence of the set of circumstances characterizing the environment in which the agents or structures interact. Focusing on racialized and gendered hierarchies within the state and its governmental institutions at the subnational level would provide an interesting perspective on how inclusion and exclusion of works; it could also provide excluded groups with a roadmap for how to design and implement effective strategies and tactics using the power available to them. In the United States, elections and campaign financing appears as a promising field for examination with significant attention focusing on the consequences of the Citizens United case adjudicated by the United States Supreme Court in 2010 that allows incorporated entities unlimited spending in electioneering communications (Lau, 2019). Using the power model to analyse the impacts of Citizens United on minority candidates and their campaigns would prove interesting indeed.

The model is designed to be flexible and adaptable to inquiries that differ from the focus within this thesis. As a framework it allows for the reduction of variables to core elements common across the many types of interactions that produce the entanglement comprising society from whichever lens applied to it. The intent of the model is for use in analysing every kind of social relation because each of these is a power relation.

References

1. Arendt H (1972) Crises of the republic. Harcourt Brace & Company, New York
2. Baldwin D (2016) Power and international relations: a conceptual approach. Princeton University Press. https://www.jstor.org/stable/j.ctt1q1xsp6.6. Accessed 7 Jan 2022
3. Berger P (1966) Identity as a problem in the sociology of knowledge. Eur J Sociol 7(1)
4. Bhargava R (1992) Individualism and social science. Clarendon Press, Oxford
5. Bilgrami A (1992) Belief and meaning. Blackwell, Oxford
6. Bourdieu P, Wacquant LJD (1992) An Invitation to reflexive sociology. University of Chicago Press, Chicago
7. Burge T (1986) Individualism and psychology. Philos Rev 95:3–45
8. Burge T (1989) Wherein is language social? In: George A (ed) Reflections on Chomsky. Blackwell, Oxford, pp 175–191
9. Cassirer E (1953) Substance and function, translated by William Curtis Swabey and Marie Collins Swabey. Dover, New York
10. Chartrand TL, Bargh JA (1999) The chameleon effect: the perception behaviour link and social interaction (PDF). J Pers Soc Psychol 76(6):893–910. https://web.archive.org/web/20110709005229/http://www.yale.edu/acmelab/articles/chartrand_bargh_1999.pdf. Accessed 6 Oct 2020
11. Currie G (1984) Individualism and global subservience. Brit J Philos Sci 35:345–358
12. Danto A (1962) Methodological individualism and methodological socialism. In: O'Neill J (ed) Modes of individualism and collectivism, 1973. Heinemann, London
13. De Graaffe N, van Apeldoorn B (2014) Corporate elite networks and US post-Cold War grand strategy from Clinton to Obama. Eur J Int Relat 20(1):29–55. https://doi.org/10.1177/1354066111433895. Accessed 27 Oct 2022
14. Domhoff G, Zweigenhaft R (2006) Diversity in the power elite. Rowman & Littlefield Publishers, New York
15. Duguit L (1917) The law and the state. Harvard Law Rev 31(1):1–185. The Harvard Law Review Association Stable. https://www.jstor.org/stable/1327671. Accessed 6 Oct 2021

16. Elster J (1989) Nuts and bolts for the social sciences. Cambridge University Press, Cambridge
17. Emirbayer M (1997) Manifesto for a relational sociology. Am J Sociol 103(2):281–317. http://links.jstor.org/sici?sici=0002-9602%28199709%29103%3A2%3C281%3AMFARS%3E2.0.CO%3B2-A. Accessed 30 Sept 2020
18. Foote N (1951) Identification as the basis for a theory of motivation. Am Sociol Rev 16
19. Foucault M (1979) Discipline and punish: the birth of the prison, translated by Alan Sheridan. Vintage Books, New York
20. Foucault M (1982) The subject and power. Crit Inq 8(4):777–795. The University of Chicago Press. http://www.jstor.org/stable/1343197. Accessed 8 June 2021
21. Foucault M (1990) The history of sexuality. In: Vol l, An introduction (trans Hurley R). Vintage Books, New York
22. Galbraith J (1983) The anatomy of power. Houghton Mifflin Company, Boston
23. Gilpin R (1983) War and change in world politics. University of Cambridge Press, Cambridge
24. Gupta A (2005) Rousseau. In: Kierkegaard's romantic legacy: two theories of the self. University of Ottawa Press. https://www.jstor.org/stable/j.ctt1ckpgbc.10. Accessed 6 Oct 2021
25. Haxton B (2001) Heraclitus: fragments. Penguin Books, New York
26. Hobbes T (1999) Leviathan. Prepared for the McMaster University Archive of the History of Economic Thought, by Rod Hay. https://socialsciences.mcmaster.ca/econ/ugcm/3ll3/hobbes/Leviathan.pdf. Accessed 29 Sept 2021
27. Jablonsky D (2001) National power. In: Cerami J, Holcomb J (eds) US army war college guide to strategy. https://apps.dtic.mil/dtic/tr/fulltext/u2/a387893.pdf. Accessed 20 Oct 2020
28. Jervis R (2017) Perception and misperception in international politics. Princeton University Press, Princeton
29. Kenyon College (2017) Turtles all the way down. https://www.kenyon.edu/news/archive/turtles-all-the-way-down/#:~:text=After%20giving%20a%20lecture%20on,turtles%20all%20the%20way%20down.%E2%80%9D. Accessed 6 Dec 2022
30. Layne C (2006) The peace of illusions: American grand strategy from 1940 to the present. Cornell University Press, Ithaca
31. Layne C (2018) The US–Chinese power shift and the end of the Pax Americana. Int Affairs 94(1):89–111. https://doi.org/10.1093/ia/iix249. Accessed 27 Oct 2022
32. Lukes S (2005) Power: a radical view. Palgrave Macmillan, New York
33. Mandelbaum M (1957) Societal laws. In: O'Neill J (ed) Modes of individualism and collectivism, 1973. Heinemann, London
34. Mann M (1986) The sources of social power. In: A history of power from the beginning to A.D. 1760, Vol 1. Cambridge University Press, Cambridge
35. Matheson C (1987) Weber and the classification of forms of legitimacy. Br J Sociol 38(2):199–215. Wiley on behalf of The London School of Economics and Political Science. https://www.jstor.org/stable/590532. Accessed 1 Feb 2019
36. Moore B (1967) Social origins of dictatorship and democracy: lord and peasant in the making of the modern world. Allen Lane, London
37. O'Connor T (1994) Emergent properties source. Am Philos Q 31(2):91–104. University of Illinois Press on behalf of the North American Philosophical Publications. https://www.jstor.org/stable/20014490. Accessed 22 Jan 2019
38. Parmar I (2017) Elites and American power in an era of anti-elitism. Macmillan Publishers Ltd 2017. https://www.academia.edu/66648556/Elites_and_American_power_in_an_era_of_anti_elitism. Accessed 5 Dec 2022
39. Peacocke C (1993) Externalist explanation. Proc Aristotelian Soc 93:203–230. http://www.jstor.org/stable/4545174. Accessed 7 Jan 2022
40. Pettit P (1993) The common mind. Oxford University Press, Oxford
41. Pizzorno A (1991) On the individualistic theory of social order. In: Bourdieu P, Coleman JS (eds) Social theory for a changing society, pp 209–244. Russell Sage Foundation, Colo, Westview, New York
42. Putnam H (1975) Mind, language, and reality. Cambridge University Press, Cambridge

43. Rousseau J (1755) A discourse upon the origin and the foundation of the inequality among mankind. International Relations and Security Network. Primary Resources in International Affairs. http://faculty.wiu.edu/M-Cole/Rousseau.pdf. Accessed 7 Oct 2021
44. Rousseau J (2017) The social contract, 1st edn. [ebook] Johnathan Bennett, pp 1–27. https://www.earlymoderntexts.com/assets/pdfs/rousseau1762book1.pdf. Accessed 6 Oct 2021
45. Searle J (1983) Intentionality: an essay in the philosophy of mind. Cambridge University Press, Cambridge
46. Spengler O (1922) The decline of the west. C.H. Beck Publishing, Munich
47. Stryker S (1987) The vitalization of symbolic interactionism. Soc Psychol Q 50:93
48. Taylor C (1992) Multiculturalism and the politics of recognition: an essay by Charles Taylor. Princeton University Press, Princeton
49. Tzu L (1963) Tao Te Ching (trans Lau D). Penguin Group, London
50. Vucetic S (2011) Anglosphere: a genealogy of racialized identity in international relations. Stanford University Press, Redwood City
51. Walt S (2018) The hell of good intentions: America's foreign policy elite and the decline of US primacy. Farrar, Straus, and Giroux, New York
52. Waltz K (2008) Realism and international politics. Routledge, New York and London
53. Warner J (2015) Rousseau's theory of human relations. In: Rousseau and the problem of human relations. Penn State University Press. https://www.jstor.org/stable/10.5325/j.ctt1wf4cqg.6. Accessed 6 Oct 2021
54. Weber M (1954) Max Weber on law in economy and society. Simon and Schuster, New York
55. Wendt A (1992) Anarchy is what states make of it: the social construction of power politics. Int Organ 46(2):391–425. The MIT Press. http://www.jstor.org/stable/2706858. Accessed 3 Aug 2020
56. Wendt A (2017) Social theory of international politics. Cambridge University Press, Cambridge
57. Wrong D (1979) Power: its forms, bases, and uses. Transaction Publishers, London and New Brunswick

Chapter 2
The Great American Identity Crisis: Manipulative Conditioned Power as Driver of Cognitive Dissonance in American Identity

> A hero ventures forth from the world of common day into a region of supernatural wonder: fabulous forces are there encountered and a decisive victory is won: the hero comes back from this mysterious adventure with the power to bestow boons on his fellow man.
>
> Campbell [5]

> Although the United States often is viewed as possessing a consensual political culture, Americans have quarrelled over what defines them as a nation. Values and symbols once shared have lost ground to new rivals, to an alternative construction of an imagined community...In a disintegrating nation-state, one manifestation of the underlying conflict over the nation's true character and mission is likely to be disagreement about the identity of one's allies and enemies.
>
> Citrin et al. [6]

Since this work views identity, interest, and the exercise of power in a causal relation and intends to examine case studies in which the United States exercises its national power, it is necessary to review the national identity of the United States, for which the concept of American exceptionalism has served as the critical element. The examination of American exceptionalism seeks (1) to establish that national identity exists, if only because of the exogenous process achieved through the looking glass self; and (2) to question how the chasm between ideology and reality created by manufactured information can damage the state and its standing within the international order. The purpose of this chapter is to establish a perspective from which the case studies might appear clearly as the focus shifts from the domestic wellspring from which its endeavours flow to the behaviour of the United States in its foreign affairs. To this end, it is necessary to review first the agential and structural forces that shape national identity in the United States.

K. L. Wedekind, *Constructing Power: A Critical and Constructivist Theory Using the American Case*, Advanced Sciences and Technologies for Security Applications,
https://doi.org/10.1007/978-3-031-84893-3_2

2.1 American Exceptionalism, Identity, and Interest

American Exceptionalism has served as the "underlying master narrative" of the United States that provides continuity between presidential administrations with different but complementary national security policy goals [28, p. 63]. Since the end of the Second World War, United States Presidents "have based their arguments for U.S. leadership on a belief in American exceptionalism…this master narrative has influenced not only presidential statements and rhetoric, but also actual foreign policy," which has been reinforced by "constructivist and liberal scholars of U.S. foreign policy [arguing] that there exists a powerful national agreement on what role the United States is supposed to play in world history because of what kind of nation the United States is believed to be" [28, p. 64]. The association between belief and action—with the former manifested by the beliefs held by the designers of foreign policy and the latter by foreign policy implementation—proves critical in that it connects national identity and interest. Within the model below, the identity formation process comprises role-taking and alter-casting. The United States takes on the role of the guarantor of the liberal international order through presidential rhetoric reinforced by the implementation of policies that align with the idea that the United States must lead within the international community "not just because of American military might or the dollar, but because the United States is exceptional" [28, p. 64]. The resulting set of beliefs proves transformational rather than transactional in that it requires the United States to exercise its power not only in its own interest, but also in the interest of others. This exogenous intersubjective social power relation transpires between the United States as an individual and the international community as a collective; subsequent exercises of power occur between the United States and the given individual entities within the international community. The corresponding endogenous aspect of the identity formation process relies upon the inherent collective character of the United States and the values that it seeks to embody in its foreign and domestic policies.

Figure 2.1 traces the process through which the sources, instrument, and forms of power that comprise the exercise of manipulative conditioned power by the United States to deceive its own people. "The executive branch can use its information control to conceal or mispresent what it is doing" to influence what the American electorate believes about itself in terms of national identity, interest, and how it exercises power [37, p. 141]. The primary line of effort depends upon personality, here closely reflecting Weber's charismatic authority, and operates through the instrument of conditioned power, eventually appearing as manipulation characterized by the presentation of an argument from a position of legitimate authority that precludes refutation by the intended power subject. The organization of government serves as a secondary source of power for additional, reinforcing arguments that constitute persuasion, which manifest as public discourse and debate of the policy position that itself constitutes manipulation. The persuasive act transpires as performative due to its inability to influence the outcome of the policy shift but does wield some degree of influence within the public that induces acceptance of the manipulation without

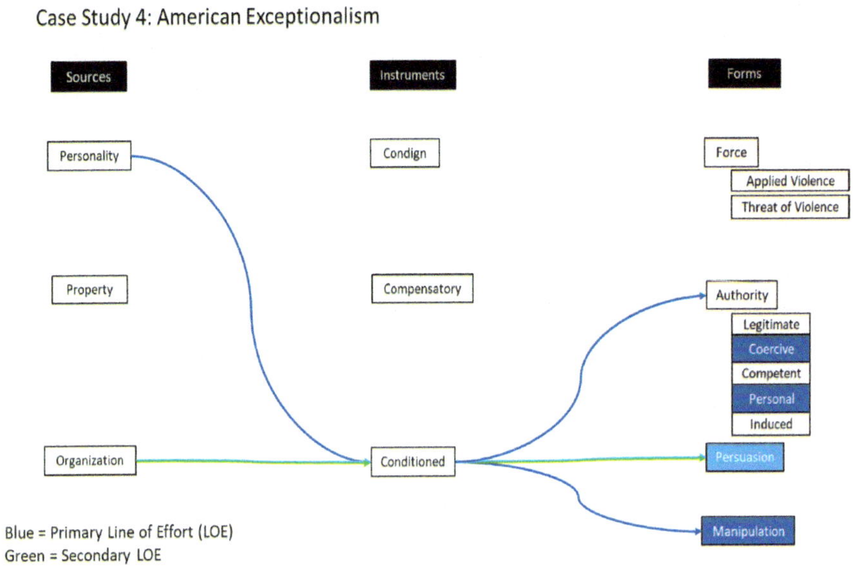

Fig. 2.1 American exceptionalism and the exercise of power (Author's own)

questioning its moral and ethical implications on the national identity and interest of the United States.

2.1.1 American Exceptionalism as the National Identity of the United States

American Exceptionalism is the manifestation of role-taking in United States social and political thought. Restad [27] argues that "American Exceptionalism is in fact a truth claim about the distinctiveness of American political and economic institutions" wherein there exist "peculiarly American approaches to government, to the economy, to culture, to religion" relative to other states (pp. 4–5). The central position arises from the proposition that "the United States—having avoided the class conflicts, revolutionary upheaval, and authoritarian governments of Europe—represented an actual example of liberty for the rest of the world to emulate. The United States was exceptional in the sense of being just a little bit better than its European cousins" [27, p. 5]. Lowry and Ponnuru [20] argue that the United States "is freer, more individualistic, more democratic, and more open and dynamic than any other nation on earth. These qualities are the bequest of our Founding and of our cultural heritage. They have always marked America as special, with a unique role and mission in the world: as a model of ordered liberty and self-government and as an exemplar of freedom and a vindicator of it, through persuasion when possible and force of arms when absolutely necessary" [27, p. 7]. Within this line of thinking resides a critical failure that appears due to insufficient criticism of American Exceptionalism, as the assumption in its truthfulness "serve[s] as a national identity feed-back loop—feeling exceptional leads to uncritically assuming one is, which leads to scholarship that takes this for granted rather than problematizing it, which again strengthens the general sense of exceptionalism." [27, p. 5, 34, p. 1031]. An inherent assumption of moral superiority resides within American Exceptionalism due to its relativist origins, with Manifest Destiny as justification for the westward expansion of the United States during the nineteenth century serving as a key example of such superiority [21, p. 24, 27, p. 6]. The act of role-taking through which the United States ascribes to itself exceptionality casts other states as somehow unexceptional; in this same way, the individuals and groups within the United States who embrace American Exceptionalism cast as others any that demonstrate a lack of fervour. American Exceptionalism therefore serves as both "a ground for objective comparison of political and economic institutions, and as an expression of an American national identity" measured by devoutness of belief [27, p. 8].

2.1.2 *The Ideals of American Exceptionalism*

The ideals, values, and aspirations comprising ideological foundation of American Exceptionalism include personal responsibility, independent initiative, economic individualism, and political voluntarism. Values encapsulated within traditional sources of moral authority such as religion, absolute moral authority, the family unit, and national pride supersede the legitimate authority of government, which meets with reluctant acceptance as a necessary evil [2, p. 54, 27, p. 5, 33, p. 182]. A common aspiration for perpetual socioeconomic improvement driven by innovation, creativity, and progress transcends social divisions, despite disparate groups offering competing solutions on how to achieve such progress [2, p. 174, 33, p. 182]. These ideals, values, and aspirations tend to produce "self-serving behaviour, atomism, and a disregard for common good" while encouraging "excessive greed and self-promotion," which indicates that Americans view the world in individualist rather than collectivist terms [19, p. 268]. This perspective includes significant disdain for globalization due to its implied association with the existence of legitimate authority beyond the United States government, against which its citizens naturally chafe. American Exceptionalism rejects notions of birth right and social or class inheritance in favour of "thinking that [individuals] create [their] own identities highlights volition and autonomy and minimizes the categorical force of race and gender in shaping our social existence" [1, pp. 420, 425–427, 33, pp. 185–186]. Being American—that is, possessing American identity or identifying as an American—in this light becomes an option that an individual can choose by believing in and practicing the tenets of American Exceptionalism, which leads to the propagation of the belief that:

> American society is exceptional...because what makes one American is not birth or ancestry but subscribing to the American Creed. The United States started from a revolution and "has defined its raison d'etre ideologically." Americans "feel emotionally connected to one another" because of their shared "ideals, values, and aspirations...This ideational foundation of America is a chief feature of American exceptionalism" [19, p. 18, 33, p. 182].

The critical aspect of American Exceptionalism derives from neither its Americanism nor its exceptionality, but the belief in the uniqueness of the American experience, especially its society and values. This reveals American Exceptionalism as naïve, if not ignorant of similar concepts of exceptionalism present in other states particularly in Western Europe [10, pp. 11–12]. In 2009, the then US President received significant rebuke after answering a question with greater empathy than what his domestic audience could tolerate: "I believe in American exceptionalism, just as I suspect the Brits believe in British exceptionalism and the Greeks believe in Greek exceptionalism" [29]. The comment revealed a change in the conception of American Exceptionalism relative to previous administrations but did not abandon the ideology. The President continued his remarks by acknowledging the interdependence between national and international interest: "I see no contradiction between believing that America has a continued extraordinary role in leading the world towards peace and prosperity and recognizing that that leadership is incumbent, depends on, our ability to create partnerships because we create partnerships because we can't solve these problems alone" [29]. The exceptionalism that the United States claims for itself neither serves as a source of power nor denies the possibility that others might possess a comparable exceptionalism reflective of the values, ideals, and aspirations that its peoples hold dearest. In this light, the uniqueness of American Exceptionalism fades. Every state and people are unique, and many started from a revolution, each has its own unique set of beliefs and credo. Despite these facts, Tyrrell argues that the American version of exceptionalism "requires something far more: a belief that the United States follows a path of history different from the laws or norms that govern other countries. That's the essence of American exceptionalism: The U.S. is not just a bigger and more powerful country—but an exception. It is the bearer of freedom and liberty, and morally superior to something called 'Europe'" [35, pp. 2–3]. A dichotomy that pits the United States against its European progenitors—and ultimately all others—creates an irreconcilable conflict in American identity that manifests via foreign policy that alters between serving as an "exemplar" and recusing itself from affairs that do not concern it or as a "missionary" that seeks to make the world safe for democracy; or rather, whichever concept the United States seeks to export through its power relations [27, p. 11]. This dichotomy manifests most evidently between the ideology of American Exceptionalism and the doctrine of America First, which suggests the existence of cognitive dissonance within the American identity.

2.1.3 *American Exceptionalism as a Gateway to Understanding American Behaviour*

The role that the tenets of American Exceptionalism play in defining how American society defines its own identity proves critical in understanding why the United States behaves as it does within the community of nations—including the belief in its own exceptionalism as license to do or not to do as it sees fit independent from any obligation incurred by its behaviour or resources. Tyrrell [35] argues that American Exceptionalism transcends the differences that set states and peoples apart as a characteristic beyond uniqueness. To achieve the exceptionalism it ascribes to itself, the United States must emancipate itself from the fate that has befallen or waits still for all other states and claim a destiny for itself unbound by history. American Exceptionalism captures this idea as the quintessential aspect of American identity for those that accept it as true. It is important to note the differences between how American Exceptionalism manifests in foreign and domestic policies. Block [3] recognizes that "the reality of American exceptionalism is that class position has always been mediated through other identities, such as ethnicity, religion, generation, and region, in shaping political behaviour" (p. 49). People in the United States experience American Exceptionalism differently based on the composition of their identity. The juxtaposition between the exclusively domestic influence of lived experiences derived from American Exceptionalism within the United States and those residing elsewhere surpass the scope of this book, though future work would do well to consider the relationship between domestic policies with respect to the power model offered in this work that includes the identity-interest formation process and exercises of power.

In 1630, a Puritan lawyer named John Winthrop delivered a sermon to the immigrants traveling from Great Britain to the Massachusetts Bay Colony before their departure titled "A Model of Christian Charity" that applied the principles of Jesus' "Sermon on the Mount" to the colonial endeavour upon which the immigrants would embark. In the conclusion of his sermon, Winthrop outlined the exemplar principle of American Exceptionalism wherein the colony established by the Puritans could serve as a model worthy of emulation by others:

> The Lord will be our God, and delight to dwell among us, as his oune people, and will command a blessing upon us in all our wayes. Soe that wee shall see much more of his wisdome, power, goodness and truthe, than formerly wee haue been acquainted with. Wee shall finde that the God of Israell is among us, when ten of us shall be able to resist a thousand of our enemies; when hee shall make us a prayse and glory that men shall say of succeeding plantations, "the Lord make it likely that of New England." For wee must consider that wee shall be as a citty upon a hill. The eies of all people are uppon us [38, p. 47].

This sermon constitutes the genesis of the beliefs that would evolve into American Exceptionalism. The identity formation process initiates in role-taking and altercasting whereby Winthrop lays claim to the role of the ideal Christian society that from his perspective should serve as an example for all other societies, which casts those as separate and therefore less virtuous. This distinction leads to the formulation of interest described in the passage above that outlines the most desirable set of circumstances offering what would become the Massachusetts Bay Colony, and eventually the Commonwealth of Massachusetts as part of the United States, the greatest chance at success once achieved. As a concept, American Exceptionalism therefore traces its origins to the religious tradition of the early British colonial period in America that precedes not only sovereignty for the United States by nearly 150 years but the establishment of Westphalian sovereignty by eighteen years. Given its genesis, American Exceptionalism serves as philosophical continuity through which the remnants of non-secular governance survives to influence contemporary political thought if Schmidt's [30] assertion that "all significant concepts of the modern theory of the state are secularized theological concepts" holds true [10, p. 15, 30, p. 36]. Winthrop could not have envisioned a secular state co-opting his words and use as justification for the interest of a secular government, but that did not prevent the progeny of his colony from adopting and applying his early formulation of identity and interest to the challenges faced in the twentieth century.

The term American Exceptionalism reappears in political rhetoric during times of tribulation in the United States, rising prominence following the end of the Second World War as a moral proposition through which the United States might convince the American people that exporting its economic prominence and military prowess offered the best promise of securing the most desirable future for their progeny. United States Presidents Kennedy and Reagan invoked Winthrop by using his term "city upon a hill" to describe the United States' role within the community of nations—notably during the Cold War—as a beacon of hope and an example worthy of emulation. Following the September 11 attacks, then United States President Bush also invoked the American Exceptionalism—this time as both the envy of the terrorists who had perpetrated the attack and as justification for embarking upon the Global War on Terrorism; the same held true for President Obama, who embraced American Exceptionalism following the transition from campaign ideology to administering [35, pp. 2–5]. Each of these episodes of presenting American Exceptionalism as justification for an action represents an exercise of persuasive conditioned power. The power holding government presents its case to the power subject constituency and argues that American Exceptionalism provides valid justification for the preferred actions of the state; further, these actions serve the national interest and therefore each citizen. The power subject receives the argument in a way that allows for a decision. The power holder must persuade the subject to select the option that the holder prefers, which in the case of American Exceptionalism has been to accept it as valid doctrine.

2.1.4 American Exceptionalism and Internationalism

During the Cold War, "ideological consensus prevailed; few then challenged the idea of a national mission to lead the 'free world' against 'communist imperialism,'" which set conditions for a singular vision of Americanism in the United States during that time; in short, the United States defined its identity relative to the Soviet Union for decades and its dissolution brought about something of an identity crisis for the United States [6, pp. 22–23]. Without the Soviet Union against which it could define its identity, the United States has sought to explain its role within the international community through various interpretations of internationalism. United States foreign policy would continue to "stress the need for continuing [United States] leadership in the world and advocate military strength as essential for defending national interests, but also recognize the need to extensive cooperation with other countries" [39, p. 428]. The absence of a traditional state actor against which the United States could define its own identity and with which it could compete removed the foil it needed to justify the version of American Exceptionalism described by Presidents Kennedy and Reagan—that of the "exemplar," the metaphorical "city upon a hill shining as a beacon for humankind." Without its antagonist, the United States identified others against which it could define a new identity by anthropomorphizing ideals as adversaries, as opposed to a specific state or organization. The United States had set a precedent for this kind of role-taking and alter-casting by declaring wars on poverty in 1964, crime in 1965, and drugs in 1971; declaring war on terror following the attacks of September 11 rather than the organizations responsible redefined the identity of the United States against a strategy as opposed to a tangible entity [4, 11, 15, 25]. The enduring nature of terrorism coupled with an identity steeped in American Exceptionalism would require the United States to pursue continuous action against any individuals or groups that employ terrorist strategies and tactics. By the time of Operation Odyssey Lightning in 2016, the case study reviewed in chapter three, the United States had conducted operations in Afghanistan, Iraq, the Philippines, Syria, Yemen, Somalia, Cameroon, Libya, and dozens of other countries in support of the Global War on Terror with the prospect of perpetual engagement looming. The open campaign for the president of the United States in 2016 offered an opportunity to reintroduce the central question surrounding the foreign policy of the United States—should the United States seek to perfect democracy at home as a beacon of hope for humanity and example for the other states, or use its power to secure the future of the liberal world order, and if the latter should the United States take such action unilaterally or through international organizations? This question considers only the two competing aspects of American Exceptionalism and neglects the possibility of other options, thereby constituting a false dichotomy in which the United States can see itself in only one of two possible ways. Maintaining the status quo established by the Global War on Terror would reaffirm the United States' vision of itself as a missionary for democracy, while retrenchment or disengagement might make the United States the exemplar that presidential rhetoric had described alternatively for decades. Another option appeared in 2016 that capitalized on a rising degree of disagreement across the American electorate about its own identity and the foreign policy of the United States, the America First policy originates from disenchantment with American Exceptionalism and its requirements, while its ascendency suggests the presence of cognitive dissonance within the national identity of the United States.

2.1.5 Fear Drives the Invocation of American Exceptionalism

The disparity between the lofty expectations set by adopting the beliefs encapsulated within American Exceptionalism and the degree of prosperity experienced by the individuals holding such beliefs creates discord when the presumed benefits fail to actualize. Disenchantment with the belief set arises when the promised or presumed prosperity it offers never appears, which compels self-reflection that leads to the assumption that the United States must have entered a period of decline as the explanation for stagnation. A period of self-perceived decline in the international stature of the United States has accompanied the political usage of American Exceptionalism. Tyrrell observes that a decline collectively perceived across a swath of American society facilitates the resurgence of the term American Exceptionalism: "When the nation was indisputably powerful and its people prosperous, Americans did not collectively require an 'ism' to serve as a guiding light…when the fates of Americans become based more on their class and less on their shared nationality, the ideological orthodoxy of American exceptionalism has emerged on a political level. A previously obscure academic term became a rallying cry for a political agenda" [35, p. 6]. Hurlburt [16] identifies the source of the resurgence in usage of American Exceptionalism as driven by common fears shared across American society: "Despite our rugged national self-image, Americans often tend toward fear: Something—whether it's American exceptionalism, child-rearing habits, continental isolation, or Internet-shortened memories—leads us to believe over and over, against all evidence, that human history has never seen anything like what our country is experiencing right now" (p. 20). Fear of decline has appeared as a recurrent theme in American public discourse for decades, Paul Kennedy's *The Rise and Fall of Great Powers* offers enduringly prescient analysis of what continues to befall the United States when stating that "the United States now runs the risk, so familiar to historians of the rise and fall of previous Great Powers, of what might roughly be called 'imperial overstretch:'…decision-makers in Washington must face the awkward and enduring fact that the sum total of the United States' global interests and obligations is nowadays far larger than the country's power to defend them all simultaneously [18, p. 515]. In Prowse [24] recognized that "declinism may be less the product of actual decline than a response to rapid economic and social change. Change is always disturbing and often perceived negatively, for the simple reason that losers tend to be more vocal than winners," offering further that socioeconomic gains in other states does not necessarily constitute a decline in the national power of the United States (pp. 1–2). This position defies the zero-sum conception of power that views a gain by one as loss by another; while the leading agent would have less relative power due to gains by another, the gain of the other does not reduce the actual or latent power of the first agent. Gains by others beginning in a significantly disadvantaged position may appear as great and swift advancements relative to those closer to their inherent carrying capacity. Economic modelling that utilizes objective data to inform predictive analytics, as opposed to the subjectivity inherent within belief. In [9], the Strategic Foresight Initiative of the Atlantic Council offered that arguments predicting a "relative economic decline of the United States" had been overstated and argued that United States national power or "global influence does not come from the size of its economic lead but rather from the proclivity and experience of the US to provide effective leadership and the ability of the international system to accept this leadership...how the United States translates political and military resources, as well as its economic strength, into power and influence will ultimately determine the longevity of its relevance on the international level" (p. 6). The key element in this observation is that power depends upon access to sources, the will to utilize instruments, and the skill in applying the forms of power, if the others comprising the international community should accept leadership from the United States either multilaterally or unilaterally.

The importance of subjective beliefs proves critical to this examination of the relationship between belief and action in the case of American Exceptionalism and US foreign policy, respectively. This proves crucial when assessing the transition away from American Exceptionalism and persuasive conditioned power towards America First and its reliance on manipulative conditioned power. Invocation of American Exceptionalism by political leaders responds to the widespread perception by the American public that the national power of the United States has begun to decline relative to other states. Relative loss of national power may occur within each of the four domains and requires a foil to serve as the other to facilitate the role-taking and alter-casting process that necessarily precedes measuring the comparable power possessed by each state.

Following 9/11, terrorism served as this key other against which the US foreign policy establishment defined the national identity of the United States through the Global War on Terror. Walt [37] concludes that "the US response to terrorism is no more impressive than the rest of its recent foreign policy. US leaders understood that terrorism was a problem back in 1993, the problem is more widespread today. Violent extremists are active in more places than ever before and with more far-reaching political consequences, often as a direct result of misguided US responses. Like other key aspects of US foreign policy, the war on terror has been a costly failure" (p. 51). These costs did not go unnoticed by the American electorate who embraced a candidate for president in 2016 that eschewed the unchallenged "conventional wisdom" of the establishment and its errors repeated across multiple theatres around the world because "where all think alike, none think very much" [37, p. 142]

Regardless of the national power domain in question—that is, whether one attempts to reconcile the military, economic, diplomatic, or informational power of the United States against the same capabilities exigent in another state—the process of measuring the national power of the United States depends upon information and requires the formation of a specific version of American identity and interest relative to the chosen other for comparison. The dependence upon information in making such determinations necessitates an exercise of power that finds its source in organization and personality—provided in this case by the United States and the presidential administration in office at a given time, respectively. Given the goal of influencing belief—which remains subjective based upon the set of dispositions constituting an identity—in either American Exceptionalism or the alternative offered by America

First, the exercise relies upon conditioned power as its instrument, which can take the form of either persuasion or manipulation to target belief and consequent behaviour. "The executive branch can use its information control to conceal or mispresent what it is doing abroad. This diminishes the ability of voters to hold officials accountable" [37, p. 141]. Belief therefore drives the usage of American Exceptionalism, not any measure indicating an actual decline of the United States. This belief exists due to the successful exercise of persuasive conditioned power in which political leaders, specifically the United States Presidents and their administrations from 1960 to 2016, have presented arguments to the American public that it accepted and thereby shaped the shared perception of American Exceptionalism as a justified, true belief. The belief in imminent decline, but not an actual decline, precedes the increased use of American Exceptionalism in political rhetoric in the United States as outlined in the preceding paragraph. The absence of an actual decline in the national power of the United States and its global standing as of 2016 necessitated the use of manipulative conditioned power to introduce and the policy known as America First to replace American Exceptionalism.

Belief in the waning power of the United States manifests as a symptom of an unclear vision of the international role of the United States caused by cognitive dissonance in the American people about their collective identity. "A poll conducted by Quinnipiac University in April 2016 found that 85% of Trump supporters agreed with the proposition that 'America has lost its identity,' many believing that the government had gone 'too far in assisting minority groups.' The electoral impact from those who feel alienated by an establishment that has ignored their economic plight is likely to continue" [8, p. 13, 12]. The association of a common perception of lost identity and an excess of assistance for others indicates an overall dissatisfaction with the status quo, including the traditional ideals outlined in policies tied to American Exceptionalism. This proves significant for several reasons: first, it affirms the existence of a relationship between national identity and foreign policy; second, it suggests that coherence plays a determining role in the relationship between national identity and the foreign policy of the United States; and third, "that the dwindling of consensus about America's international role follows from the waning of agreement on what it means to be an American" [8, p. 8]. Disagreement within the electorate emanating from an identity crisis amounting to cognitive dissonance in American identity and any consequent concept of interest, which set conditions that allowed an alternative to American Exceptionalism gain prominence through the election of Donald Trump as President of the United States in 2016.

2.2 The Art of Manipulation

In 2016, the Republican Party in the United States released a document titled "America Resurgent" that outlined its policy platform, of which American Exceptionalism forms the foundation:

> We believe that American exceptionalism—the notion that our ideas and principles as a nation give us a unique place of moral leadership in the world—requires the United States to retake its natural position as leader of the free world. Tyranny and injustice thrive when America is weakened. The oppressed have no greater ally than a confident and determined United States, backed by the strongest military on the planet. Quite simply, the Republican Party is committed to rebuilding the U.S. military into the strongest on earth, with vast superiority over any other nation or group of nations in the world. We face a dangerous world, and we believe in a resurgent America [26].

The document further outlines a platform focused on national security achieved through maintaining United States "military superiority" through increased defence spending, collaborating with allies and working through cooperative security mechanisms, and acting against adversaries when necessary "to advance the cause of liberty and [United States] national security" without "pander[ing] to world opinion and neglect[ing] the national interest" (GOP 2016). According to the platform by designing it as such the Republican Party sought to "affirm [the party's] tradition of world leadership" through "enormous power—and the prudence to use it sparingly, precisely, and only in grave necessity" where involvement would supersede intervention with a "[requirement] for consultation, not permission to act" while "[leading] from the front—and ensur[ing] all others do their parts as well" (GOP 2016). The platform "embraces American exceptionalism and rejects the false prophets of decline and diminution. It is…the way we will lead the world into a new century of greater peace and prosperity—another American Century" (GOP 2016).

Three details in the platform prove noteworthy. First, the description of the desired state of the United States military as having achieved "superiority" as opposed to "supremacy." United States military doctrine defines the former as "that degree of control by one force that permits the conduct of its operations at a given time and place without prohibitive interference" and the latter as "that degree of control wherein the opposing force is incapable of effective interference" [36]. This shift reduces the threshold of acceptance for military prowess by the United States relative to its competitors and indicating a national security strategy that seeks retrenchment following fifteen years of continuous conflict waged as part of the Global War on Terrorism, it is important to note that the United States reopened a theatre of operations in Libya through Operation Odyssey Lightning during the same time as when the GOP wrote its America Resurgent platform. Despite the bombastic tone of the document, it sets a goal for the military at a lesser degree than the total domination connoted by supremacy and to which the document otherwise alludes by affirming the right of the United States to act unilaterally. Second, the platform identifies the United States national interest as a singular concept—just as this book does—and affirms the idea that the national interest of each state incorporates all aspects of national power comprising each domain and all its aspects into a singular concept. Third, the use of the first-person plural in the closing sentence of the preceding paragraph refers to the United States conditionally and presupposes Republican victory in the 2016 United States Presidential Election. Despite the victory of the Republican candidate in that election, the GOP platform here described has been supplanted by an alternative approach to foreign policy. The act of displacing the GOP platform, which embraces the traditional interventionist perspective founded upon the idea of American Exceptionalism, with the principle of America First based in nativist isolationism constitutes an exercise of manipulative conditioned power. A reconciliation between the 2016 GOP America Resurgent policy platform and the 2017 National Security Strategy reveals that the latter makes no mention of American Exceptionalism and outlines a strategy counter to it. A transition occurs between campaigning and governing that mirrors a similar transition between persuasive and manipulative conditioned power catalysed by the obtainment of legitimate authority for the elected official and party. Once elected, the president does not need to convince through persuasion but rather can decree by making statements that find their position elevated regardless of truthfulness, which creates the opportunity to manipulate. The presentation of beliefs from a position of legitimate authority denies those captive to it with the same perception of choice as that afforded when the same argument emanates from any of the other forms of authority. Manipulation denies a choice to the reluctant power subject; while making statements does not make them true, issuing them with legitimate authority does make them more consequential for any subsequent actions taken in pursuit of the ideals or goals outlined within such statements. In this way, the combination of legitimate authority and the use of conditioned power constitutes manipulation as the form of power exercised.

Since 1945 the United States has based its foreign policy doctrine on "the belief that the United States has a special mission to redeem the world by extending liberty and democracy to all peoples" [8, p. 1]. This messianic version of American Exceptionalism has necessitated that the consequent actions undertaken to implement the foreign policy of the United States during this time reflect an internationalist character. "This belief in the necessity for American leadership flows in part from the recognition that powerful states sometimes need to cajole others into cooperating in order to achieve common goals. If the world's most powerful country disengaged completely and did not encourage other states to address global problems, selfish national interests might loom larger and achievable solutions to challenges such as climate change might never be reached [37, p. 58]. All the while beneath this internationalist veneer a competing perspective on foreign policy and action has been ruminating within the United States among groups that viewed the internationalist approach as the manifestation of "reckless visions of regional or global hegemony—especially in the Middle East -[that left] the home front to languish and the nation open to ridicule" [8, p. 1]. This competing perspective on foreign policy in the United States favoured prioritizing its own national interest ahead of all other considerations, including preservation of the liberal international order that facilitated the prosperity enjoyed within the United States since the end of the Second World War, the decade of Pax Americana from 1991 to 2001, and the time since September 11, 2001 during which the United States has led a Global War on Terrorism. The underlying sentiment that enabled the transition away from American Exceptionalism emanates from a loss of belief in the internationalist mission of the United States and its cost, no matter how measured and despite that accepting "such a future would mean a rejection of all that lies at the heart of their national identity" [35, p. 17]. Changes in belief reflect changes in identity and interest. Rejection of American Exceptionalism includes the rejection of its virtue, which had been that the United States serves its own national interest best by exercising its national power in a way that accounts for its singular interest as well as the collective interest of the community of states, most significantly by ensuring the stability and the continuation of the liberal international order—in favour of policies indifferent to the collective interest of the community of states that amount to a rebirth of isolationism in the form of America First, enabled by the perceived failures of domestic and foreign policy in action such as the 2007–2008 recession, continued degradation of race relations, and the protracted conflict required by the Global War on Terrorism. Where American Exceptionalism intends to act as both in the interest of the United States and the collective, America First views the exercise of power as either serving the United States or the other; what had been reconciled previously now proves irreconcilable to the detriment of all. America First offers individuals disappointed with their lot in life a way of justifying their sentiments. Beneath this observation reverberates the sentiment that human beings have a need to understand themselves and the world in which they live; "We hold the beliefs that we do not only to understand our environment, but also to meet our personal and social needs…people need to live comfortably with themselves and this strongly influences their beliefs" [17, p. lxxiv].

The 2017 National Security Strategy defines the meaning of America First. The strategy begins with a letter from the United States President to the American people that asserts a new and different vision of the United States and its place in the world. The letter presents perspective as fact, which suggests that it could constitute an exercise of manipulative conditioned power due to the denial of choice to the intended audience. The legitimate authority wielded by the President of the United States in issuing the letter and strategy denies choice to the American people, thereby constituting manipulation [40, pp. 28–29]. Deviating from the principles outlined in Republican Party's platform after having lost the popular vote in the election reinforce the proposition that the vision for America First described in the 2017 National Security Strategy and foreign policy actions taken under its guise constitute an exercise of power via manipulation. The strategy sacrifices American Exceptionalism's self-appointed "special mission" of the United States to exercise its power supporting the liberal international order in favour of serving its own interests. The mutual exclusivity of American Exceptionalism and America First appears in statements that declare "We are prioritizing the interests of our citizens and protecting our sovereign rights as a nation. America is leading again on the world stage" that misrepresents how a state must act to lead international efforts that depend upon collaboration and trust [32, p. 1]. "Threat inflation," a trend that Walt [37] identifies as a tactic used by the US foreign policy establishment to perpetuate the perception of clear and present existential threats against the United States that justify its exercises of power, follows as the 2017 National Security Strategy explains that "the United States faces an extraordinarily dangerous world" with an international community consisting of only enemies: unpredictable, nuclear-armed states, terrorist organizations holding territory to establish hostile governance; transnational criminal organizations driven by the drug trade and human trafficking; and allies that exploit the United States economically and militarily" [32, p. 1, 37, p. 138]. The argument then presupposes the consequence of these alleged circumstances and projects them upon the American people while declaring a break from the status quo: "Too many Americans had lost trust in our government, faith in our future, and confidence in our values. Nearly one year later, although serious challenges remain, we are charting a new and very different course [32, p. 1]. The manipulation becomes clear as the letter to the American people disavows the legacy of American Exceptionalism by claiming that America First will have "restored confidence in America's purpose…recommitted ourselves to our founding principles and to the values that have made our families, communities, and society so successful…we will serve the American people and uphold their right to a government that prioritizes their security, their prosperity, and their interests. This National Security Strategy puts America First" [32, p. 2]. To implement this vision, the United States President "withdrew the United States from the Trans-Pacific Partnership (TPP), the trade pact including a dozen countries that represent some 40% of the world economy" that escalated into a trade war with China, "but the United States decision not to join the pact it had done so much to bring about has instead relieved pressure on China to reform while penalizing U.S. exporters seeking to sell to the new pact's members" [14]. Withdrawal continued as the United States vacated the "2015 Paris Climate Accord, the 2015 Iran Nuclear Deal, the Intermediate-Range Nuclear Forces Treaty (INF), UNESCO and the U.N. Human Rights Council, the World Health Organization, notably amidst a global pandemic, and the Open Skies Treaty" [14].

Despite its domestic application, this manipulation forces consequences upon the stability of the liberal international order. The United States' "allies will need to be more finely attuned to the anger and frustration pouring out of Middle America, along with the repercussions they have on Washington's self-confidence and capacity" [35, p. 18]. "Without a United States to rely on, allies are faced with the unappealing choice of deferring to a powerful neighbour, building up their own military capabilities, or both. Either outcome works against United States interests and can create opportunities for competitors, such as China, to gain standing at our expense" [14]. The United States fails to serve its own interest by divorcing it from the collective because "the Trump world view pits nationalism against America's international responsibilities" [8, p. 10]. Parmar observes that President Trump's.

> Personal style—racialized, disrespectful, aggressive—and his dismissal of the concerns of other states under the banner of 'America First' undermines American credibility. The moral authority—to the degree that it exists—at the heart of the liberal hegemonic project has little or no role to play in the embrace of the 'principled realism' of Trump's 'America First' national security strategy [23, p. 242].

Where American Exceptionalism allowed for the possibility of reconciling the United States' interest with others including the collective interest of the international community, America First proves irreconcilable with any interest beyond the United States; that is, American Exceptionalism allows for exercises of power that serve both the United States and others, while exercises of power taken under the guise of America First serve either the United States or the other—thereby embracing the false dichotomy of the zero-sum perspective on power.

The legitimate authority of the office of the President combines with matter-of-fact rhetoric to deny the social aspect of persuasion wherein an aspiring power holder affords the intended power subject with choice to accept or deny the arguments presented to convince the subject to modify his, her, or their behaviour and instead constitutes an act of manipulation through conditioned power. The magnitude of potential consequences caused by this manipulation can be viewed in terms of extensiveness, comprehensiveness, and intensiveness. In the direct sense of the power relation, the manipulation transpires between the United States President as the power holder and the American people as the power subject; the policies enacted under the authority of the President necessarily transpose upon the American people because the policies of the President serve as the directive for action for the United States. In an indirect sense, domestic power relations in the United States impact its relations with all other states, whether through bilateral or multilateral engagement, and by extension their peoples. While these states and people are not party to the power relation in question, its disposition bears upon them. The extensiveness of the domestic power relation in the United States therefore proves universal; manipulation in this power relation subjects nearly all, if not all, to the consequences associated with the United States abdicating its traditional vision of itself and role in the world. The comprehensiveness of the manipulation juxtaposes its acute and systemic impacts. The information domain provides the medium within which the power relation transpires and transcends the other three national power domains; the legitimate authority of the power holder enables the manipulation by denying a choice to the power subject when supplanting the traditional national and bipartisan responsibilities to the global community with its abandonment in favour of an inwardly focused policy characterized by nativism and isolationism void of responsibility beyond the self. Acute comprehensiveness manifests as tacit acquiescence that reflects both ignorance and apathy such that the power holder proceeds unchallenged; systemic comprehensiveness appears in the series of withdrawals from international agreements, exacerbated by combative rhetoric against allies and domestic opponents comprising the loyal opposition as well as advocating for the domestic use of condign power and the military that constitutes force at least in its non-violent, psychological form. While acute comprehensiveness may be slight, systemic comprehensiveness effects both domestic and foreign entities directly and indirectly. The intensiveness of the manipulation proves singular; the power holder denies the subject with any semblance of choice by wielding its legitimate authority to invoke will. The power subject possesses no means of equitable recourse within the information domain. The application of manipulation through the information supported by legitimate authority compels the power subject to seek recourse via other instruments, forms, and sources of power.

The United States withdrawal from international agreements in pursuit of an America First foreign policy constitutes disengagement. Abdicating global responsibilities proves counter to the national interest of the United States by preventing it from working through the negotiating structures established by each abandoned agreement and treaty. The United States forces itself to attempt to influence outcomes as an outsider to the liberal international order that it designed with its allies and which facilitated its ascent during the second half of the twentieth century. Diminishing its own influence allows for others—whether state or non-state actors—to exert greater power in support of its own interest. The connectedness of the globalized world magnifies the consequence as global challenges such as state violations of international law, violent non-state actors, and climate change compel collective action. By pursuing a foreign policy of withdrawal, the United States has hindered its ability to achieve its own national interest and fulfill its moral, ethical, and legal obligations to the global community. The United States and Americans as individuals prove entirely unexceptional; but while human nature defies exceptionalism due to universality, the United States has the potential—or rather, exigent, latent power—that could become exceptional under certain conditions and within specific parameters. Denying American Exceptionalism facilitates the transition from aligning the interest of the United States with the community of states to a self-serving policy summarized by the phrase America First. To achieve some measure of exceptionalism, the United States would need to align its national interest with the interest of the collective—the global community—in perpetuity. This same principle holds true for any state and any people: exceptionalism—if it can be achieved rather than only striven towards—requires a sustained transformation from inherent selfishness to learned selflessness. If it remains true that "the true author of the subjugation of a people is not so much the immediate agent, as the power that permits it having the means to prevent it," the United States will bear the burden of responsibility for failing to exercise its national power to preserve the liberal international order, the modicum of stability that it provides, and its own national interest [14, 31, p. 39]. Jervis [17] recognizes the challenge inherent within this proposition: "Even without the compounding influence of nationalism, leaders and members of the mass public tend to be self-righteous and to have difficulty understanding the hopes and grievances of others, with unfortunate consequences for international relations" (p. xxvi). The conflict between the actions of the United States and its vision of itself indicate cognitive dissonance in the American psyche that manifests as the great American identity crisis.

2.2.1 Cognitive Dissonance in American National Identity

The contest between American Exceptionalism and America First created cognitive dissonance within the national identity of the United States due to the mutual exclusivity of these ideologies. "The consequence has been growing ideological polarization, political stalemate, and the rise of new political currents, such as populist authoritarianism and socialism, that were previously restricted to the margins of politics in the United States" [3, p. 54]. The United States cannot view itself as exceptional—as either exemplar or missionary—by divorcing its national interest from the global interest. By placing itself first, the United States forsakes either role where it could serve as a leader in favour of the part of the antagonist against the liberal international order that it created to serve both its national interest and set conditions for stability and progress globally. By adopting the America First policy, the United States takes on the role of antagonist by exalting its own national interest, which divorces it from any common interest while also casting the external agent and their interest as a competing other.

It is critical to note that both American Exceptionalism and America First serve as justifications for exercises of power by the United States. As ideologies, a presidential administration could invoke either to justify the same action or inaction; that is, the purpose of the rhetoric is to convince through persuasion or manipulation that whatever activity selected by the administration and influenced by US foreign policy establishment is appropriate regardless of the policy that action serves. The greatest difference between American Exceptionalism and America First appears in its presentation as the core of either position can be invoked to justify any action even if it contradicts precedent. The message remains the same—that "Right is only in question between equals in power" and the United States will "do what it can, and [others] suffer what they must" [31, p. 22]. Identity is therefore contingent upon belief, which is shaped through interaction with the other and the structural environment within which the interaction transpires. Identity and interest remain subjective and transient, relative to the world in which one lives and in response to the inherited social structures and strictures that form the ties that bind.

2.3 Conclusion: The Echoes of Manipulation

What Horace intended when he wrote "dulce et decorum est pro patria mori" turned to dust as he did, though the impact of his idea lived and thrived among those industries of government that depend upon able-bodied, military aged youths of sound mind and their willingness to fight and die for an intangible organization, the manifestation of an idea that exists only in the abstract as visible through the actions that execute policy [7]. Wilfred Owen dispelled such beliefs when he outed the maxim as a lie in his poem of the same name albeit thousands of years later because by then and to him, any truth that those words had held once, was gone [22]. An American corollary comes from George Patton when he explained that "no bastard ever won a war by dying for his country. He won it by making the other poor bastard die for his country" [13, p. 64]. By reversing the logic, Patton kills the romanticism and eliminates the manipulative power of Horace's words. Perhaps Horace believed in the worthiness of dying for the Empire and if he did then it would not constitute an attempted manipulation, but rather an act of persuasion intended as encouragement, Horace presents the argument and leaves it to his audience to decide for themselves in its virtue. If an entity were to invoke Horace's words to inspire—that is, motivate one to self-compel—youths to volunteer to serve in the military and this entity did so because it served its own purposes, such as to fill the ranks simply for the purpose of creating or perpetuating a source of power, and lacked a justified true belief in the maxim then the act of appealing to a romanticized classicism because of its resonating effect on the intended power subject would comprise an attempted exercise of manipulative conditioned power. Success would depend upon the outcome, which in this case would be whether the recipient enlisted—the motivation for which emanates from inspiration planted by belief in service to one's country. Underlying this effort to motivate lies a more sinister purpose as foreign policy actions require individuals and organizations to carry out the orders. "Vested interests within the government and the broader foreign policy establishment have significant advantages in shaping what the public knows about international politics and foreign policy, and these advantages tilt the competition among policy alternatives in their favour. In other words, the marketplace of ideas is rigged" [37, p. 138].

Horace's words echoed across generations for thousands of years from Rome in the time of Augustus to the First and Second World Wars two thousand years later and beyond, inspiring young folks into undertaking an endeavour that they otherwise might not have because they believed in the propriety of doing so despite it being counter to their interest. Arriving at such a belief can only occur through an exercise of manipulation. The power holder presents an idea to the unwitting subject that the perpetuation of an idea or organization surpasses their own lives in importance, and that sacrificing themselves affords the best way to sustain the idea or organization. While sacrifice does prove necessary, the sacrifice of life does not. The sacrifice of interest does. The act of placing the interest of another ahead of one's own offers a pathway to exceptionalism.

The irreconcilability of international and domestic interest therefore has caused the crisis of conscience in the collective American psyche that appears in the two competing visions of itself; in the first, the United States defines its interest singularly and separate from the community of nations as described by America First, while in the second the United States aligns its interest with its neighbours' so that one reinforces the other. The globalized world allows for states to overcome the zero-sum notion that has characterized foreign policy for centuries. A gain by one does not necessarily cause loss by another. Diversity in sources, types, and instruments of power increase the ways in which power holding states and power subject states can engage with one another. The overall power relation becomes much less important than the multitude of nuanced engagements transpiring simultaneously and concurrently within each domain of national power. Describing one state as dominating another portrays neither an accurate nor precise representation of a given power relation, consequently providing no useful insight for those seeking to understand how and why states behave.

References

1. Appleby J (1992) Recovering America's historic diversity: beyond exceptionalism. J Am Hist 79(2):419–431. https://doi.org/10.2307/2080033
2. Baker W (2005) America's crisis of values: reality and perception. Princeton University Press
3. Block F (2019) The end of American exceptionalism: the social question in the United States. In: Breman J, Harris K, Lee CK, van der Linden M (eds) The social question in the twenty-first century: a global view. University of California Press. https://www.jstor.org/stable/j.ctvr7fcnz.8. Accessed 8 Dec 2020

4. Bush G (2001) Transcript of President Bush's address to a joint session of Congress on Thursday night, September 20, 2001. http://edition.cnn.com/2001/US/09/20/gen.bush.transcript/. Accessed 11 Dec 2020
5. Campbell J (1968) The hero with a thousand faces. Princeton, Princeton University Press, p 53
6. Citrin J, Haas E, Muste C, Reingold B (1994) Is American nationalism changing? Implications for foreign policy. Int Stud Q 38(1):26–27. https://www.jstor.org/stable/2600870?read-now=1&seq=23#page_scan_tab_contents. Accessed 7 Jan 2022
7. Conington J (1882) The Odes and Carmen Saeculare of Horace. George Bell and Sons, London. http://www.perseus.tufts.edu/hopper/text?doc=Perseus%3Atext%3A1999.02.0025%3Abook%3D3%3Apoem%3D2. Accessed 15 Dec 2020
8. Curran J (2018) "Americanism, not globalism": President Trump and the American mission. Lowy Institute for International Policy. https://www.jstor.org/stable/resrep19793. Accessed 7 Jan 2022
9. Dadush U, Garrett B, Ignatius D, Steinberg J (2012) The United States and the global future stable. Atlantic Council. Strategic Foresight Initiative. http://www.jstor.com/stable/resrep03367. Accessed 7 Jan 2022
10. De Boever A (2019) Aesthetic and political exceptionalism. University of Minnesota Press. https://www.jstor.org/stable/10.5749/j.ctvm202pr.4. Accessed 8 Dec 2020
11. Drug Policy Alliance (2020) A brief history of the drug war. https://drugpolicy.org/issues/brief-history-drug-war. Accessed 11 Dec 2020
12. Fukuyama F (2016) American political decay or renewal: the meaning of the 2016 election. Foreign Affairs, July/August 2016. https://www.foreignaffairs.com/articles/united-states/2016-06-13/americanpolitical-decay-or-renewal. Accessed 15 Dec 2020
13. Gavin J (1958) War and peace in the space age. Harper and Brothers, New York
14. Haass R (2020) Trump's foreign policy doctrine? The withdrawal doctrine. The Washington Post; May 27, 2020. https://www.washingtonpost.com/opinions/2020/05/27/trumps-foreign-policy-doctrine-withdrawal-doctrine/. Accessed 7 Jan 2022
15. Hinton E (2015) Why we should reconsider the war on crime. Time. March 20, 2015. https://time.com/3746059/war-on-crime-history/. Accessed 11 Dec 2020
16. Hurlburt H (2020) The only thing we have to fear is fear-based politics: the politics of crisis policymaking. In: Democracy beyond COVID-19. New America. https://www.jstor.org/stable/resrep25415.7. Accessed 8 Dec 2020
17. Jervis R (2017) Perception and misperception in international politics. Princeton University Press, Princeton
18. Kennedy P (1988) The rise and fall of great powers. Unwin Hyman, London
19. Lipset M (1996) American exceptionalism: a double edged sword. W.W. Norton and Company, London
20. Lowry R, Ponnuru R (2010) An exceptional debate: the Obama administration's assault on American identity. Nat Rev Online. https://www.nationalreview.com/magazine/2010/03/08/exceptional-debate/. Accessed 9 Dec 2020
21. Merk F, Merk L (1963) Manifest destiny and mission in American history: a reinterpretation. Harvard University Press, Cambridge, MA
22. Owen W (1920) Poems. Chatto and Windus, London
23. Parmar I (2019) Global power shifts, diversity, and hierarchy in international politics. Ethics Int Aff 33(2):231–244. https://doi.org/10.1017/S0892679419000091. Accessed 22 June 2021
24. Prowse M (1992) Is America in decline? Harvard Bus Rev. https://hbr.org/1992/07/is-america-in-decline. Accessed 25 Nov 2024
25. Rector R (2014) The war on poverty: 50 years of failure. The Heritage Foundation, September 23, 2014. https://www.heritage.org/marriage-and-family/commentary/the-war-poverty-50-years-failure. Accessed 11 Dec 2020
26. Republican Party of the United States (2016) 2016 republican party platform. https://www.presidency.ucsb.edu/documents/2016-republican-party-platform. Accessed 7 Jan 2022

27. Restad H (2011) The past in political science: American exceptionalism and U.S. Foreign Policy. In: Norwegian Institute of International Affairs. Paper presented to the ECPR conference, Reykjavik, Iceland, August 2011. https://ecpr.eu/filestore/paperproposal/b794a228-2972-4fc7-ae16-3b309e417416.pdf. Accessed 9 Dec 2020
28. Restad H (2019) Whither the "City upon a hill"? Donald Trump, America first, and American exceptionalism. Texas Nat Secur Rev 3(1) (Winter 2019/2020). Print: ISSN 2576-1021. Online: ISSN 2576-1153. https://tnsr.org/2019/12/whither-the-city-upon-a-hill-donald-trump-america-first-and-american-exceptionalism/. Accessed 8 Dec 2020
29. Scherer M (2009) "Obama too is an American exceptionalist," Swampland blog, April 4, 2009, Times Magazine Online. http://swampland.blogs.time.com/2009/04/04/obama-too-is-an-american-exceptionalist/. Accessed 9 Dec 2020
30. Schmitt C (1985) Political theology: four chapters on the concept of sovereignty (trans Schwab G). MIT Press, Cambridge, Massachusetts
31. Strassler R (1998) The landmark thucydides. Free Press; Touchstone ed. Edition
32. The White House (2017) The national security strategy of the United States: a new national strategy for a new era. https://trumpwhitehouse.archives.gov/articles/new-national-security-strategy-new-era/#:~:text=The%20publication%20of%20the%20National%20Security%20Strategy%20%28NSS%29,security%20vision%20into%20practice%20on%20behalf%20of%C2%A0fellow%20citizens. Accessed 7 Jan 2022
33. Thomson I (2010) Culture wars and enduring American dilemmas; culture, class, and American exceptionalism. University of Michigan Press, Ann Arbor, MI. https://www.jstor.org/stable/j.ctt22p7hg8.10. Accessed 7 Jan 2022
34. Tyrrell I (1991) American exceptionalism in an age of international history. Am Hist Rev 96(4). https://www.jstor.org/stable/2164993. Accessed 7 Jan 2022
35. Tyrrell I (2016) American exceptionalism, from Stalin with love. Aeon; October 10, 2016. https://aeon.co/ideas/american-exceptionalism-from-stalin-with-love. Accessed 7 Jan 2022
36. United States Air Force (2017) Doctrine advisory: control of the air. https://www.doctrine.af.mil/Portals/61/documents/doctrine_updates/du_17_01.pdf?ver=2017-09-17-113839-373. Accessed 7 Jan 2022
37. Walt S (2018) The hell of good intentions: America's foreign policy elite and the decline of US primacy. Farrar, Straus, and Giroux, New York
38. Winthrop J (1630) A modell of Christian charity. Collections of the Massachusetts historical society (Boston, 1838), 3rd series 7:31–48. Hanover Historical Texts Collection. Scanned by Monica Banas, August 1996. https://history.hanover.edu/texts/winthmod.html. Accessed 15 Dec 2020
39. Wittkopf E (1986) On the foreign policy beliefs of the American people: a critique and some evidence. Int Stud Q 30(4):425–445 (21 pages). Wiley on behalf of The International Studies Association. https://doi.org/10.2307/2600643; https://www.jstor.org/stable/2600643. Accessed 7 Jan 2022
40. Wrong D (1979) Power: its forms, bases, and uses. Transaction Publishers, London and New Brunswick

Chapter 3
Military Power in Operation Odyssey Lightning

> It's a faceless enemy, unrecognizable, blending in with hundreds of others. It is everywhere. In cafés, in the alleys of the Casbah, or in the very streets of the European quarter.
>
> Colonel Mathieu, *The Battle of Algiers*, 1966

> There are two types of people in this world: us, and them.
>
> US Marine Officer, Operation Odyssey Lightning, 2016.

This case study examines the military operation executed by the United States against the Islamic State of Iraq and Syria in Libya (ISIS-L) in 2016 within the vicinity of Sirte, Libya, which the United States termed Operation Odyssey Lightning. While the military domain of national power frames the chapter, its analysis focuses on how the identity and interest formation process at both the national-strategic and theatre-tactical levels that transpired between the parties to the conflict, including not only the United States and ISIS-L but also the Libyan factions vying for authority in an ongoing civil war. Analysis of the order of battle concentrates on the ways in which each belligerent cultivates and applies power through the sources, instruments, and forms reviewed in the preceding chapters to determine how the actions taken constitute an exercise of power in terms of intentionality and effectiveness. The purpose of this chapter is to analyse what happened in Operation Odyssey Lightning by applying the power theory to demonstrate its validity and utility. This case study is intended to demonstrate the identity and interest formation process of the United States relative to ISIS-L and its consequent exercise of condign power in the form of military force utilizing physical and psychological violence.

The juxtaposition between the United States and ISIS-L and their respective exercises of power in this episode highlights the role that identity and interest play in determining the actions that an agent might undertake. Each collective agent defines its identity and interest relative to the other such that the identity and interest of the United States exists uniquely within this case study relative to ISIS-L, and vice

K. L. Wedekind, *Constructing Power: A Critical and Constructivist Theory Using the American Case*, Advanced Sciences and Technologies for Security Applications,
https://doi.org/10.1007/978-3-031-84893-3_3

versa. Just as time and space limit the engagement from August to December 2016 and within the vicinity of Sirte, Libya, the identity and corresponding interest of each agent exists only within this same time and space; subsequent engagements between the United States and ISIS-L produce new identities and interests for each party. The sources, instruments, and forms of power available to each agent to project or resist the other exist contemporaneously, which facilitates analysis of how each utilized its power and the accompanying degree of success in achieving its interest within the engagement. The United States and ISIS-L both draw upon the same types of sources, instruments, and forms of power, and both project and resist the overtures of the other simultaneously until the episode concludes.

As the engagement transpires within the military domain, this case study focuses primarily on the association between property, condign power, and force. Property in the form of military materiel and corresponding capability, enabled by organization as a secondary source of power, serves as the key source of power for the United States. Condign power "wins submission by the ability to impose an alternative to the preferences of the individual or group that is sufficiently unpleasant or painful so that these preferences are abandoned," and can transpire as the application or threat of physical violence or psychological non-violent force [15, p. 4, 40, p. 24]. The application of condign power bears an association with coercive authority, which accounts for the capability and willingness to use force—whether in the actual application or threat of either violence targeting the physical body or psychological faculties of the opposing agent. In this case study, only the United States exercises condign power and force due to the sources of its power coming primarily from property and secondarily from organization. Throughout the duration of Operation Odyssey Lightning, United States military units utilize airpower to exercise condign power by delivering force in its violent physical form; ISIS-L possesses no comparable capability that might afford it with the means to resist and inevitably succumbs to the United States military units through which the national power of the United States manifests in this instance.

A secondary line of effort through which the United States exercises power within this case study emanates from property and organization as sources of power applied through conditioned power in the form of persuasion. The effect of condign power and the use of force demonstrates to both parties and all others observing the ramifications of engaging the United States in similar environments to that of Operation Odyssey Lightning. Whether explicit or implicit, implied or inferred, the military actions by both sides directly shape the perceptions of observers in ways that reflect the degree of their success; that is, the result of an exercise of power conditions the beliefs not only of the agents party to it but also those agents that observe the conflict despite their position as uninvolved. The consequence of an action persuades the observer to perceive of the victor as strong and the defeated as weak. The binary effect of observation proves superficial and represents a false dichotomy; while it may be the case that the victor is strong and the defeated is weak, other explanations exist for why an episode favoured one agent over the other and these explanations emanate from the constitution of the agent including identity, interest, and power and the degree to which these three elements align. The concept of bimodal symmetry explains

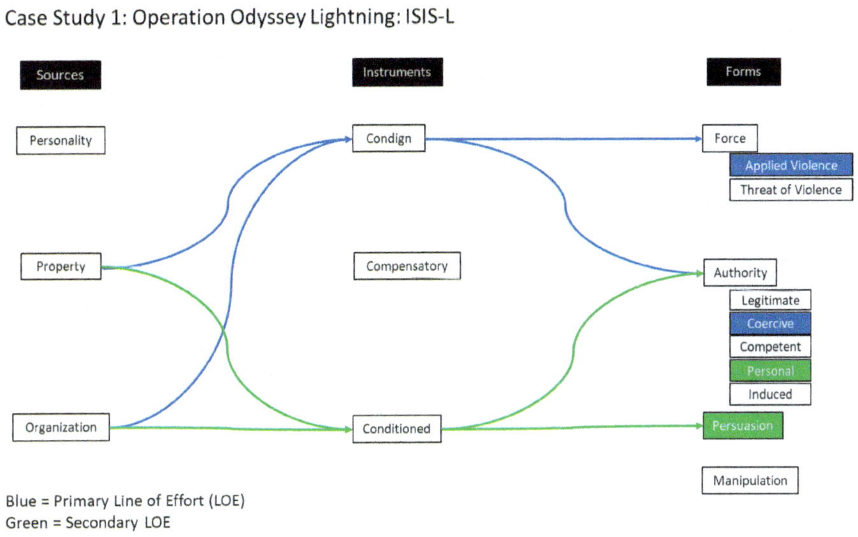

Fig. 3.1 US exercise of power within Operation Odyssey Lightning (Author's own)

how these properties emerge within the United States military units that executed Operation Odyssey Lightning.

Figure 3.1 traces the process through which the sources, instruments, and forms of power used by the US military during Operation Odyssey Lightning. Property and organization as sources of power consolidate into condign power within the military unit that then projects force by applying violence and coercive authority. Property and organization also allow the military unit to exercise conditioned power by persuading any observing agents to modify their behaviour to avoid the circumstances that characterized the conflict between the US military and ISIS-L in 2016 in the vicinity of Sirte.

3.1 Identity and Interest Formation Through Role-taking and Alter-Casting

The identity and interest formation process for the agents that participated in Operation Odyssey Lightning began at least as early as March 2011 with the adoption of United Nations (UN) Security Council Resolution (SCR) 1973, which demanded an immediate cease fire to ongoing internal conflict in Libya, the establishment of a no-fly zone, and Libyan compliance with international humanitarian and human rights law; Resolution 1973 also granted UN member states the authority to use military force to protect Libyan civilians and populated areas [33]. This justification for the use of military force against Libya in 2011 finds its basis in the Responsibility to Protect doctrine, which demands the protection of human life and rights even in instances that might violate the territorial integrity of states; in this case, the human rights of Libyan citizens superseded the sovereignty of the Libyan state [13, p xii]. UN SCR 1973 defined the identities of the parties to the conflict. By passing UN SCR 1973 and authorizing the use of military force against Libya, the UN Security Council condemned Libya by casting it as the other by distinguishing between it and the UN member states cooperating to an acceptable degree with international norms at the time; the passage of UN SCR 1973 also reaffirmed the role of the UN Security Council as arbiter of international conflict by exercising the legitimate authority that the community of states invested in it to identify, condemn, and authorize action to enforce norms of behaviour on states. Resistance against the impending military action becomes the primary national interest for Libya with the goal of survival as a state. The use of military force against a sovereign state deemed irresponsible, reckless, and a threat to regional and international stability by the community of states—such as the case of Libya in 2011 that drove the passage of UN SCR 1973 and the actions executed in support of it—remains a legitimate and realistic consequence for any state acting outside of these norms. An interesting question for future research about the collateral effect of UN SCR 1973 would ask to what degree state sovereignty became conditional upon cooperation with international norms of behaviour.

The passage of UN SCR 1973 cast Libya as an outlier relative to the community of states and set the two against each other, exemplifying a relation between an individual and collective, respectively. Reducing the collective, in this instance representing the entire community of states except Libya the outsider, transpired upon the initiation of military operations undertaken by the United States, in addition to at least 10 NATO member states and four Arab League member states [35]. These states reduce the collective to identifiable individuals with separate identities, roles in the conflict, and interests in its outcome; while each are worthy of consideration, this work focuses on only one. The United States outlined its interest in acting against Libya as seeking "to prevent a humanitarian catastrophe and address the threat posed to international peace and security by the crisis in Libya…[and] to support an international coalition as it takes all necessary measures to enforce the terms of UN SCR 1973" [32, p. 1]. This statement clarifies the identity and corresponding interest for the United States in undertaking its actions in 2011 against a marginalized Libya as an enforcer of international norms as outlined in this instance by UN SCR 1973 as a way of mitigating a humanitarian crisis and a threat to global stability through cooperative military action with states like itself. The corresponding military actions demonstrated a high degree of intentionality, though the degree of effectiveness proved inversely low and did not achieve the desired end state—quite to the contrary, the military actions undertaken in 2011 would precipitate the onset of instability leading to civil war in Libya and subsequently requiring additional military action in 2016 with a slew of new actors with unique identities, interests, and power seeking to replace the defunct Libyan state.

The 2011 Libyan intervention eliminated the government and along with the identity and interest that it had provided to Libya until its dissolution, which also removed any mechanism for implementing governance within Libyan borders. A

National Transition Council existed for nearly one year, but dissatisfaction with its progress in re-establishing the rule of law in Libya led to protests and subsequent dissolution in 2012 [25]. An organization known at the General National Congress (GNC), comprising freely and fairly elected officials, succeeded the National Transition Council but failed to accomplish the goal of its mandate to produce a constitution acceptable to the various movements and militias jockeying for dominance over the others. Violence erupted in 2014 when the deadline passed for delivery of the GNC constitution. With no such document in place the GNC dissolved which required a new election [39, p. 2]. The subsequent elections created a new organization, the House of Representatives (HoR), which sought to protect its new mandate by aligning itself with the Libyan National Army (LNA) in eastern Libya when the GNC refused to dissolve. In December 2015, an agreement brokered by the United Nations known as the Libyan Political Agreement outlined in UN Resolution 2259, the GNC relinquished its position and ceased to exist [34]. In its place, the Government of National Accord (GNA) became the internationally recognized governance entity in Libya. Despite the arrangement endorsed by the United Nations and its unanimous support from the members of the Security Council, the HoR and LNA soon rejected the GNA and its mandate [2]. Since then, these two rival organizations have sought to re-unify Libya under their authority and remain engaged in an ongoing, albeit low-intensity, civil war. As distinct entities, both the GNA and LNA possess unique identities, but comparable interests as both seek to rule Libya. This suggests that intersubjective social-power relations wherein the two agents claim the same interest but have different identities inherently trend towards conflict as a way of reconciling their competing interest.

Organization as a source of power reveals itself in the turmoil that transpired in Libya between 2011 and 2016. The establishment of organizations fit enough to act as such implies the completion of the identity-interest formation process due to the role-taking and alter-casting required during the organizing process. The evolution of governance between 2011 and 2016 in Libya proves noteworthy due to its influence on the identities and interests of the actors present in 2016. As an organization forms, it necessarily binds agents as members through an ideology. The short duration of organizations such as the NTC and GNC indicate that the members did not coalesce with a sufficient degree of bimodal symmetry, defined as the ratio of belief in terms of intensity, extensiveness, and comprehensiveness across the membership body of an organization in the ideals of the organization; conversely, that the GNA and LNA have persisted for so long in a fractured Libyan state indicates a high degree of bimodal symmetry in each. Within Libya, both the GNA and LNA ascribe for themselves and each other the same roles; that is, each organization views itself as the rightful legitimate authority in Libya and the other as a pretender. Conflict ensues in such cases of mirrored role-taking and alter-casting.

3.2 New and Constant Identities and Interests

The instability created by the military intervention in 2011 set conditions for continuous political turmoil and supporting military engagements between rival militias that would allow a non-state actor, the Islamic State in Iraq and Syria (ISIS), to gain a foothold in Sirte, Libya. The situation, exacerbated by a migrant crisis, proved untenable and prompted a second military intervention in Libya in 2016 at the request of the GNA [19]. Led by the United States and the United Kingdom with support from France, Italy, and several states neighbouring Libya, the operation would target and destroy ISIS in Libya (ISIS-L). While performed at the request of the GNA in support of their ground forces, the United States directly targeted ISIS-L [8]. The United States would reprise its role from the 2011 episode by exercising condign power by using military force, including both physical and psychological violence, to intervene in Libyan affairs under the auspices of seeking to mitigate a humanitarian crisis but primarily focused on eliminating the foothold established by ISIS-L in Sirte, Libya. By taking on its same role and pursuing its same interest from the 2011 intervention, the United States serves as a constant in terms of identity and interest. From the United States perspective, the only difference between 2011 and 2016 came in the form of the adversary; instead of targeting the legitimate Libyan government, the governance vacuum created by having supported the overthrow of the Libyan government in 2011 set conditions for a stalemate civil war between two evenly matched agents in the GNA and LNA; the governance vacuum also allowed for ISIS-L to consolidate in Sirte and exert its control over the city. The presence of an agent that neither the GNA nor LNA would tolerate cast an identity upon ISIS-L as an unwelcome interloper that required annihilation. "Truly 99% of the population and almost all of the major militia factions were anti-ISIS and viewed the group as a foreign incursion on Libyan territory. Most were open to embracing a shared effort against the group" [26, p. 157].

Within the context of Operation Odyssey Lightning, the exogenous identity of ISIS-L derives from its designation as a terrorist organization by the United States in May 2016. By designated ISIS-L as a terrorist organization, the United States casts ISIS-L as the other whose noncompliance with international norms has exceeded the limit of acceptability; such alter-casting sets conditions that will allow the United States to exercise condign power through the use of force, specifically with its military and a combination of physical and psychological violence. The United States claims a two-part role for itself by designating ISIS-L as a terrorist organization. Making such designations presumes that an actor possesses the authority to do so, but as a member of the hierarchy-less community of states no single state holds the legitimate authority to do so within the international order; the United States uses its legitimate authority as a sovereign state to establish a policy that views ISIS-L as a terrorist organization from its perspective alone. Beyond the limits of its sovereignty, the United States must exercise one of the other four types of authority. Since part of the terrorist designation serves as a moral condemnation, the United States relies upon its personal authority, which finds its source in prestige that emanates from national character and standing within the community of states. As Parmar notes, prestige proves critical to the ability "To maintain hegemony and secure favourable outcomes [as] much depends on actual and perceived American global behaviour, and its 'image', not to mention high levels of legitimacy at home in regard to its global mission" [27, p. 248]. Prestige, or the perception of prestige, can legitimize personal authority by conditioning others to accept the corresponding actions of the United States as norms, whether new or reinforcing sustained actions.

The endogenous identity of ISIS-L derives from the Salafi-jihadist ideology of its parent organization, the Islamic State of Iraq and Syria, that seeks to bind state and religious governance under sharia law using condign power in the form of physical and psychological violence [17, p. 12]. The identity formation process proves central to ISIS efforts to establish itself as a legitimate sovereign state as opposed to an upstart non-state actor by embracing an ideology that depends on violence and cultivating a membership willing to use violence—both against external agents and its own members—to maintain a sufficient degree of bimodal symmetry that allows the organization to project power in the same way as a sovereign state does. Part of "ISIS ideology holds that Islam must be cleansed from within and that Muslims who do not strictly follow their ISIS-authorized interpretation of Islam should be declared apostates and put to death" while another aspect outlined in "'*Aqidah wa Manhaj al-Dawlah alIslamiah fi al-Takfir*' (Islamic State Creed and Methodology of Takfir, or excommunication) states that whoever embraces democracy and secularism, and all governments that do not rule by the sharia law, are deemed *kafir* or apostate, which legitimizes their killing" [17, p. 13]. The willingness to use violence on its actual members and potential pools of recruits suggests a high degree of bimodal symmetry within ISIS and accounts for one of its most prominent identity characteristics, which enables it to project power, control territory, and establish governance. While ISIS governance depends upon condign power, it has succeeded by offering relative stability in areas devastated by degradation of pre-existing governance like Syria and Libya [17, pp. 22–23]. In exchange for the restoration of basic social and economic services, ISIS implements a social contract that requires complete submission to its ideology and materiel support from those living under its rule, which constitutes not legitimate but coercive authority [28]. For those living in places where the pre-existing governance apparatus has failed or been destroyed view the choice as one between the chaos left by the vacuum or submission to ISIS in exchange for a semblance of society, such as it is under an organization that depends upon the instrument of condign power as a source of power, as opposed to personality, property, or organization. The opportunity afforded by the destruction of the Libyan government in 2011 set conditions that allowed "the Islamic State [to enter] Libya to take full advantage of the chaos after the revolt of 2011. As in Syria, the group rejected the option of cooperating with other Salafi groups and instead tried to impose its own agenda and organization. The Islamic State's enemy was as much other Islamist movements as the competing government forces" [16, p. 46]. This rejection of other organizations with similar purposes distinguishes the unique identity that ISIS-L claims for itself while casting comparable organizations as others that possess either different beliefs or believe the same but with less devotion; in either case, belief or its stringency separates like organizations, which shapes the identity of the organization and the way in which it behaves as a collective entity. Identity formation through role-taking and alter-casting plays a critical role in ISIS efforts to gain and control territory. ISIS members take on the role of judge, jury, and executioner through their willingness to use violence against any individuals that do not submit to their interpretation of Islam and sharia law; individuals or groups that embrace different beliefs or behave less devoutly become cast as others, which ISIS-L views as justification to exercise condign power using force through the application of physical and psychological violence—an act upon which ISIS depends to sustain itself. In areas where ISIS can control territory, one either joins, complies, or dies [17, p. 23].

3.3 Operation Odyssey Lightning

The United States designated the 2016 military action against ISIS-L and in support of the GNA as Operation Odyssey Lightning. The United States relied upon its property and organization as sources to exercise condign power using force in the form of physical and psychological violence demonstrating the correlation between property, condign power, and force, mirroring the exact way in which its adversary, ISIS-L, exercised its power in seeking to establish itself as the legitimate authority in Sirte. Major differences arise in the property available to the belligerents and the way in which each organizes itself; the United States finds its center of gravity in its military, specifically the ability of tactical units to task organize and sustain operations, which depend upon organization and property as sources of power, respectively. Despite a similar reliance on its military, ISIS-L depends upon condign power and force as sources rather than as instruments and forms of power. This difference in access to the sources of power suggests that the duration of military operations undertaken by the United States far exceeds the comparable capability of ISIS-L due its possession of relatively limited property that diminishes the magnitude of its organization into a viable supply chain from point of origin into Sirte. The consequence of less access to property as a source of power correlates to a reduced operating capability, which in this case appears as the ability of ISIS-L fighters in Sirte to withstand the sustained airpower campaign that the United States undertakes; the use of the term fighters to describe the ISIS-L contingent in Sirte in 2016 as opposed to units reflects the degree of organization that a dearth of property conveys upon the individuals seeking to coalesce into a collective. Property therefore bears a direct correlation upon organization; greater possession of property inspires a greater degree of organization, though this correlation proves value-neutral in that the simple possession of vast amounts of property in a consequently well-organized collective does not indicate probability or likelihood of success in the endeavours undertaken by such a collective. Access to property does impact operating capabilities and inspires greater organization but neither property nor organization determine the outcome of an engagement—the exercise of power is the sole arbiter of the result.

The United States conducted Operation Odyssey Lightning primarily through the application of force in the form of airstrikes. Operating under the mandate to minimize collateral damage and kill no civilians, the United States relied exclusively on guided munitions. With no ground forces from the United States available to designate targets, remotely piloted aircraft provided the only way of guiding munitions on target. This required a near-constant presence of remotely piloted aircraft flying over Sirte that could be used to surveil potential targets and designate them when piloted aircraft were available. When piloted aircraft were unavailable, armed remotely piloted aircraft were used to designate targets and deliver munitions. During hours of darkness when aircraft were off station, destroyers escorting the amphibious ships fired illumination rounds from their five-inch guns to disrupt the ability of ISIS-L forces to reinforce their positions. Conducting the Operation in this way required organizing the supporting US military units in a way that integrated their respective capabilities in time and across space. Creating periods of time when the effects of weapons could support the GNA advance relied upon the synchronization of aircraft, including supporting aircraft like tankers to refuel fixed wing aircraft in addition to rotary wing aircraft that could requisition maintenance and ordnance supplies from land bases while the amphibious shipping remained near enough to Sirte to facilitate adequate time on station for aircraft. The land-based remotely piloted aircraft flew from bases in neighbouring states but were piloted by United States Air Force personnel physically located in the United States [6]. Accomplishing the mission depended upon synchronizing multiple airframe assets based in disparate and austere locations above Sirte in alignment with the movement of GNA forces, with which none of the United States forces had any form of verbal communications [29]. In total, aircraft from the ARGMEU flying in support of Operation Odyssey Lightning flew 764 sorties for 1266 flight hours, dropped 290 bombs, and fired 164 missiles that destroyed 25 pieces of equipment, 197 defensive positions, and killed nearly 1000 ISIS-L fighters; to support combat operations, supporting aircraft flew 196 sorties, transported 1452 personnel, and delivered 78 tons of cargo to and from the amphibious ships while DDGs fired 285 illumination rounds from its five-inch gun [12, 22, 30]. Remotely piloted aircraft conducted an additional 300 strikes in support of the operation [23]. These statistics indicate the presence of organization as a source of power for the United States at the tactical level as multiple, disparate units organized complementary capabilities into an exercise of condign power using force and the application of violence [37]. The US military units involved did not possess the organic capabilities required to conduct the Operation successfully without additional resources so their ability to align their physical assets and integrate across their respective service cultures proved critical to the success of the United States in its campaign against ISIS-L.

3.4 Property, Condign Power, and Force

The violence wrought by the United States during Operation Odyssey Lightning serves as a reminder of the effectiveness of condign power in both its physical and psychological forms to reset the exigent conditions in favour of the power holder. The physical violence associated with the use of military force and application of national condign power demonstrates the inverse correlation between intensiveness and comprehensiveness as opposed to extensiveness. During the operation within its geographically defined boundaries, the force applied by the United States depended upon the high degree of intensity afforded by the form of physical violence delivered by AV-8B Harriers and AH-1W Cobra helicopters. These platforms, supported by sea and land bases, delivered munitions against ISIS-L and provided the United States forces with few limits in choosing when and how to apply force during the operation. While the comprehensiveness of the operation proves measurable in a single domain, its dominance within it achieves total supremacy in that ISIS-L possessed no means of counteracting or defending itself against the way in which the United States exercised its power. The limits of the operation in time and space also limit the extensiveness of the effects of force and physical violence, in that the impacts of violence affect only those present within the boundaries and scope of the operation.

If an inverse correlation can be said to exist between intensiveness, comprehensiveness, and extensiveness, then the reciprocal appears true for the psychological, or non-physical, effects of force in the form of violence. The duration of the operation and the area where it occurred limit the intensiveness and comprehensiveness of the associated psychological violence, as opposed to the extensiveness of the same which impacted not only the ISIS-L fighters in Sirte but also the other manifestations of the organization across North Africa and the Middle East. The psychological impact of the United States' use of force would bear upon warring Libyan factions, as well. While the elimination of ISIS-L from Sirte would have bolstered morale—certainly among the GNA and likely the LNA—it would not have gone unnoticed that the United States could just as easily begin targeting other actors within the region.

The purpose of force in the form of violence is to kill enough members of an opposing entity such that those who remain can no longer endure and the battle ceases. The results of physical violence reveal both the power holder and subject within the military domain and demonstrate the effectiveness of force and condign power in terms of intensity and comprehensiveness, albeit limited in terms of extensiveness. The extensiveness of the acute effects of physical violence remains limited by the capabilities of the tool employed, which indicates that the immediate impact of force in the form of physical violence achieves relatively less extensiveness than does the psychological impact of the violence itself. The injuries inflicted upon physical bodies pales in comparison to the psychological impacts upon those who observe the violence or learn of it second-hand. This distinguishes the condign from the conditioned power inherent within the use of force. A single exercise of force in the form of physical violence produces two effects—the physical harm produced upon the intended target and the psychological impact upon those who share aspects of an identity with those who have been hurt, which manifests emotionally as sympathy and through reciprocal exercises of power albeit not necessarily in the same violent form. The use of violence may achieve an immediate and decisive effect, but remains limited by the constraints of time, space, and resources; its application brings consequences for those who do use it at the hands of those who do not succumb to its effects. Sustaining the use of force and application of violence directly correlates to the degree of compensatory power, and by extension possession of property, controlled by the power holder.

During Operation Odyssey Lightning, the United States demonstrated the correlation between property as a source of power, the instrument of condign power, and the use of military force as an exercise of national power. The possession of property in terms of funds available for appropriation to military purposes, including research and development, equipment procurement, training, and operations, enables the projection of military power. For the United States, investments in research and development facilitate the purchasing and subsequent use of military equipment that it uses to sustain its global presence and project power using military force. The critical component of this exercise in compensatory power comes not from the purchase of assets, but the ability to support the operating costs for their near continuous use. The primary United States military unit—comprising three surface vessels with 28 aircraft, 4500 sailors and marines, and dozens of vehicles embarked known as an Amphibious Readiness Group-Marine Expeditionary Unit (ARGMEU)—that executed Operation Odyssey Lightning exemplifies this relationship between property, condign power, and force. The operating cost for each ARGMEU, of which the United States maintains seven in various states of readiness, exceeds $1.2 billion annually [7, pp. 61–63]. The ARGMEU provides the United States with an ever-ready capability to deliver military power across the world in accordance with its interest. For this service, the United States pays $3.2 million each day per ARGMEU team, or $2.3 billion to $4.6 billion annually—the equivalent of 5–10% of the entire defence budget for the United Kingdom in 2019, which is the highest defence spending in Europe [5]. Figures 3.2 and 3.3 illustrate the property required for a single ARGMEU and demonstrate the significance of property as a source of military power. It is critical to note that the mere possession of property does not maintain any type of association with military success, as indicated by many recent conflicts in which non-state actors have resisted military actions undertaken against them by nation-states. The possession of property in terms of materiel and funding does correlate to an ability to project power, understood as the capability to engage across time and space in accordance with the interest of the agent in question.

This image illustrates the materiel composition of the Amphibious Readiness Group, which is part of the property that served as a source of power for the United States during Operation Odyssey Lightning [7, p. 62]. The availability of materiel, including not only its existence but maintenance, limits capabilities that translate into the projection or resistance of an exercise of power. Materiel therefore constitutes a source of power within the category of property. The financial resources available to the United States enable two critical capabilities that sustain its military operations by

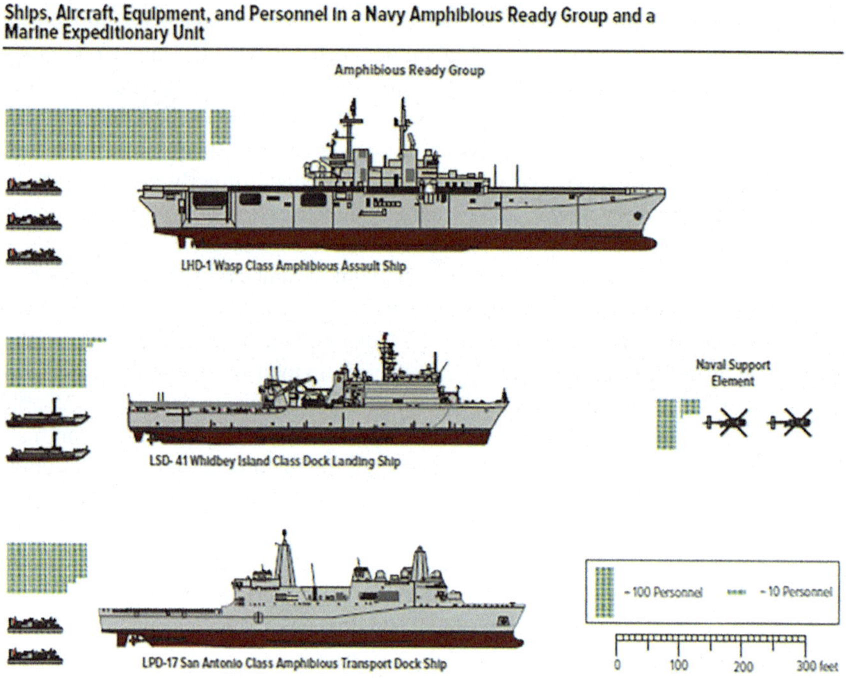

Fig. 3.2 Ships, Aircraft, Equipment, and Personnel in a Navy Amphibious Ready Group and a Marine Expeditionary Unit

sea and air for the ARG and MEU, respectively. The ability to conduct replenishment-at-sea (RAS) and underway replenishment (UNREP) allows for the Navy vessels comprising the ARG to receive fuel, sustenance, and maintenance parts such that the limiting factor for remaining at sea becomes the crew itself rather than the ships and their components. To receive these critical supplies, the ARG ships depend upon auxiliary shipping known as replenishment tankers, whose sole mission is to sustain supply lines to deployed vessels that serve as sea-based assets carrying latent or projecting kinetic power abroad. During its 2016 deployment that included Operation Odyssey Lightning, the ARG would complete 44 replenishments-at-sea across 87,000 nautical miles [36]. This capability allowed the ARG ships to remain on station off the coast of Libya to serve as the sea-base for the piloted aircraft from the MEU that delivered ordnance in support of GNA ground forces, thereby giving them tactical advantage and facilitating the destruction of ISIS-L in Sirte.

This image illustrates the materiel of the Marine Expeditionary Unit, which is part of the property that served as a source of power for the United States during Operation Odyssey Lightning [7, p. 63]. Each MEU deploys with an organic sustainment capability provided by its complement of 12 MV-22 Ospreys, which are tiltrotor, assault support aircraft that take off and land vertically like a traditional helicopter but fly like

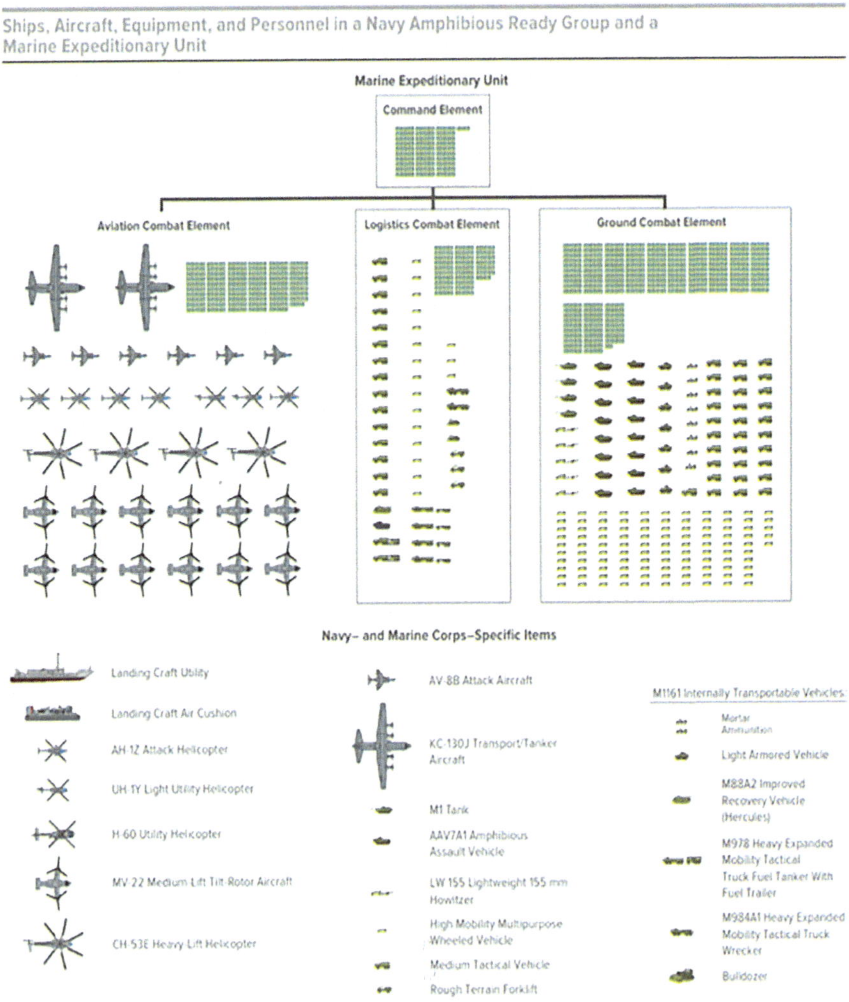

Fig. 3.3 Table of equipment for a Marine Expeditionary Unit

a turbo-propeller airplane [21]. Each MV-22 can transport up to 55,000 pounds internally, 10,000 pounds sling-loaded externally, or 24 combat troops, and can conduct aerial refuelling to increase its range to 2100 nautical miles with one refuelling [21]. During Operation Odyssey Lightning, MV-22 Ospreys flew 196 sorties for 259 flight hours, including the movement of 1452 individuals and 155,645 pounds of cargo that proved critical to sustaining combat operations against ISIS-L fighters in Sirte [22]. This airbridge provided the ARGMEU with the ability to remain engaged against its enemy instead of requiring it to break contact for periodic port calls to rearm. This capability demonstrates the value of property and organization as sources of power for sustaining military operations from sea-based assets or in austere environments;

such a capability can only exist where an agent possesses sufficient property in the form of materiel and organizes it into lines of operation with corresponding lines of sustainment.

Property as a source of power provides both the replenishment-at-sea and MV-22 airbridge capabilities that sustain combat operations in austere environments for the United States. The possession and use of property depend upon the amount of compensatory property available to a given actor; thus, a direct correlation between compensatory power and property exists. Compensatory power that remains in its latent stage grants the power holder with access—understood as the potential to gain property—while actualized compensatory power procures property that the holder uses to project power upon its subject. The possession of property therefore creates the possibility of the existence of a capability, the capacity of which depends upon the scale of access to the property; that is, property alone does not constitute capability, but it does create the possibility that a capability can exist. The organization of property in its employment determines what capabilities exist and what remains simply materiel. Viewing organization and property as complementary sources of power demonstrates the complexity of the origins of power and the inherent interdependence between its sources. It is critical to note that neither property nor organization correlate to success in the outcome of the intersubjective social power relation within which an entity might use either or the two combined. The experiences of the United States military throughout operations in Iraq and Afghanistan since 2001, along with many other case studies through military history and especially those considered counterinsurgencies, demonstrate that the amount of property and degree of organization bears no determinative effect on the outcome of an engagement between two entities [9, p. 1].

3.5 Not so Near Peer Competition

Property insufficiently organized reduces the overall effectiveness of the agent seeking to exercise power. The Russian Federation demonstrated this concept coincidentally during Operation Odyssey Lightning by sailing its only ship capable of launching and recovering fixed wing aircraft, the *Admiral Kuznetsov*, to the Syrian coast to conduct operations. Within three weeks of its entrance into the Syrian conflict, two embarked aircraft, one SU-33 Flanker and one MiG-29K, were lost due to malfunctions aboard the vessel during either take-off or landing. Consequently, Russia moved its embarked aircraft from the *Kuznetsov* to an airbase in Syria [14]. While the United States maintained an operational tempo of two to ten sorties per day for 100 consecutive days from one of its ten amphibious ships capable of launching and recovering fixed wing aircraft, the Russian Federation failed to project power from its only ship with similar capabilities for more than 21 days. This comparison between the American and Russian capabilities excludes the 11 Nimitz and Ford class aircraft carriers that the United States operates in addition to its amphibious ships; each aircraft carrier holds upwards of 60 fixed wing aircraft, primarily variants of the

FA-18 Hornet, in addition to several dozen supporting rotary week aircraft. The sortie generation capacity of one United States amphibious ship equates to 12–15% of one aircraft carrier [11]. Given that Russia proved unable to sustain air operations from its aircraft carrier, it seems that no reasonably comparable capability can be said to exist in the Russian navy. With no exigent capability, no corresponding capacity can exist and therefore no comparison can be made between the United States and Russia in terms of projecting air power from sea-based assets. The point of the comparison between the United States and Russian capabilities in projecting military power from sea-based assets is that property alone does not enable the exercise of condign power through the use of force, at least for exercises of power within the military domain.

3.6 The Missing Phases of Operation Odyssey Lightning

Upon conclusion in December 2016, the United States had used military force, specifically condign power in the form of physical violence, to destroy ISIS-L in Sirte, which allowed the GNA to seize the town so that it might establish itself as the legitimate authority for governance. Failure to undertake stabilization operations or enable the civil authority indicates that "the United States sees Libya as a purely military problem and ignores the political dimension" [10]. Since then, the GNA has established a municipal government and military force responsible for its protection known as the Sirte Protection Force, although questions regarding the effectiveness of either remain as indicated by the absence of any Libyan national government leaders from the second anniversary of the liberation of Sirte, to include those representing the area, as well as their inability to support either the municipal government or the Sirte Protection Force with materiel required to protect the city and clear remnants of war from it, which would make it safe for reconstruction and reconstitution [1, 3]. Pack [26] observes that "The fall of Sirte [in December 2016 did] not represent the defeat of ISIS, nor end the threat of other jihadist groups hijacking Libya's post-Qadhafi transition, nor [did] the removal of ISIS from Sirte solve the city's concerns over governance, security, and marginalization. Jihadist groups have been woven into Libya's post-Qadhafi pattern of appeasement, deputization of militias, and statelessness. Only attempts to combat those issues, rather than simply to push ISIS out of specific territories, would be treating the root cause of the problem" (p. 151). Operation Odyssey Lightning created an opportunity to treat the source of the problem, but the United States limited its scope to the application of force using condign power via property and organization. The decision to forgo stabilization operations ashore limited the overall effects of the United States exercise of power. As of July 2021, reconstruction efforts in and around Sirte remain in their infancy, while the LNA—backed by France, Egypt, and the United Arab Emirates—has begun conducting airstrikes of its own against critical infrastructure in Sirte as the United States recommenced airstrikes of its own against a resurgent ISIS-L in areas south of the city [1, 4]. While the intensity of the conflict has changed since Operation Odyssey Lightning, Sirte and the surrounding area remains a contested area characterized by the continuous threat of violence and its occasional realization by no fewer than four entities and as many as eight, between the GNA, LNA, ISIS-L, and the United States in addition to France, the United Arab Emirates, Egypt, and the United Kingdom, respectively [4].

The failure by the United States to initiate stabilization operations and subsequently enable civil authority—at least in Sirte and its immediate vicinity—provides one possible explanation for the continuation of conflict and degraded governance in Libya and, if valid, warrants questioning whether Operation Odyssey Lightning constitutes an exercise of national power, specifically in terms of intentionality and effectiveness. The dependent factor in this determination proves the original intent of the operation. If the United States set out to destroy ISIS-L, then the operation cannot constitute an exercise of power because ISIS-L persists within Libyan territory thus failing to meet the intent to destroy; if the focus of the operation were limited further to ISIS-L in Sirte, then an exercise of power may have transpired but the fact that conflict continues and governance remains degraded within Sirte indicates that the exercise itself has proved ineffective, which makes it not an exercise of power at all.

The decision to forgo stabilization operations and enable civil authorities in Libya deviates from precedents set by the United States in Operations Enduring and Iraqi Freedom. In both conflicts, the United States and coalition forces dedicated resources towards Human Terrain Teams and Provincial Reconstruction Teams that focused on identifying and developing individuals to serve within new governance bodies and critical infrastructure that could provide services capable of sustaining the local society; no such efforts focused on cultivating liberal governance in Sirte or Libya that might afford economic, social, and political opportunities to the populace while withstanding attempts by political organization incompatible with the contemporary international order were undertaken by the United States or any other member state from the North Atlantic Treaty Organization or European Union [24, p. 14–15]. Pack [26] emphasizes the opportunity created by successful collaboration at the tactical level only to allow a lack of strategy squander the gains:

> Operation Odyssey Lightning was the first major success of the Obama administration's war against ISIS and the one in which collaboration with local Arab forces had worked most effectively. (Nearly all of the Department of Defense's militarily successful partners in Iraq and Syria were Kurds.) And yet, the fight against ISIS in Sirte was a wasted golden opportunity to unite Libya. Policymakers had chosen to solve the immediate problem in a way that was convenient for the short term rather than to harness it as an opportunity to address systemic, longer-term concerns (p. 156).

This reluctance towards reconstituting a failed state, despite its role in setting determinative conditions, establishes a new precedent wherein the United States views its interest in terms of itself alone as opposed to its interest in alignment with the collective interests of each state and its respective contribution to global stability; that is, Operation Odyssey Lightning marks the beginning of a transition for the way in which the United States defines its global responsibilities, separating its own singular interest from that of the collective community of states comprising the international order.

3.7 Conclusion

The role-taking and alter-casting that transpired between the United States, Libya, and the international community across the decades preceding the conflict between the United States and ISIS-L in Sirte in 2016 marginalized Libya as an outlier within the community of states and set conditions that allowed for both the failure of Libya as a state and ISIS-L to establish a foothold within its territory that necessitated further international intervention in the form of military action due to the incompatibility of ISIS-L ideologies and consequent governance with the national interests of the neighbouring NATO members. Identity played a critical role in separating Libya from the international community as an outlier just as ISIS-L has remained a non-state actor despite having possessed many attributes of a state for years. Recognizing or denying the identity claimed by an agent produces an effect upon the interest of the agent. Within the context of Operation Odyssey Lightning, Libya seeks to re-establish governance across the territory that it held prior to 2011 and the United States exercised its power to support this goal by destroying ISIS-L in Sirte. Libya and the shared vision of Libyan identity—held by both the GNA and LNA despite their ongoing conflict—receives affirmation through this support. The role that Libya has taken for itself and that cast upon it by the international community align; the role that ISIS-L has taken contrasts with the identity that NATO states cast upon it. The result is an alignment of national interest between a fractured Libya and its neighbouring NATO states against ISIS-L. The United States, as the only NATO member both capable and willing to exercise its military power in support of an idea of Libya represented by the GNA and LNA, justifies its actions as part of the Global War on Terrorism and executes its appointed mission of eliminating ISIS-L in Sirte. Identity therefore begets interest, which therefore determines the exercise of power.

The quote from the US Marine officer at the beginning of this chapter echoes the same sentiment expressed by Kipling [20] nearly one hundred years prior when he stated: "All the people like us are We/And every one else is They" [38, p. 128]. This distinction reveals the relativity inherent within the identity formation process. The interaction between two agents causes each to determine their identity relative to the other within the engagement. Each must decide whether the other is part of however they define "we" or "they." Jervis [18] recognizes that "Once a person develops an image of the other—especially a hostile image of the other—ambiguous and even discrepant information will be assimilated into that image…people perceive what they expect to be present" (p. 68). An agent with the perspective of the Marine officer reflecting a world in which only "us and them" exist will find ample evidence to support their claim. The problem is that there are many different peoples and type of people within each group. The worldview of either friend or foe only is a false dichotomy because it denies possibilities that exist self-evidently in a world of pluralism. This same logic that applies to the individual as agent applies to the state as agent. Jervis extrapolates his observation on the individual to the state: "If they think that a state is hostile, behaviour that others might see as neutral or friendly will be ignored, distorted, or seen as attempted duplicity. This cognitive rigidity reinforces the consequences of international anarchy" [18, p. 68]. Perception shapes belief about self and other, which is an act that constitutes one way in which role-taking and alter-casting occurs. From this identity, an interest arises that determines how the agent should exercise its power.

References

1. Alharathy S (2019) Haftar drones target Al-Qardabiya Base in Sirte. The Libya Observer; October 5, 2019. https://www.libyaobserver.ly/inbrief/haftar-drones-target-al-qardabiya-base-sirte. Accessed 7 Jan 2022
2. Al-Warfalli A (2016) Libya's eastern parliament votes against U.N.-backed government in Tripoli. Reuters. https://www.reuters.com/article/us-libya-security-un-idUSKCN10X1DY. Accessed 7 Jan 2022
3. Assad A (2018) Sirte celebrates second anniversary of liberation from ISIS. The Libya Observer; December 17, 2018. https://www.libyaobserver.ly/news/sirte-celebrates-second-anniversary-liberation-isis. Accessed 7 Jan 2022
4. Assad A (2019) Al-Sarraj and British chargé d'affaires discuss developments in Libya. The Libya Observer; October 13, 2019. https://www.libyaobserver.ly/news/al-sarraj-and-british-charg%C3%A9-daffaires-discuss-developments-libya. Accessed 7 Jan 2022
5. Chuter A (2019) UK's top diplomat urges decisive defence budget increase. Defense News; May 14, 2019. https://www.defensenews.com/global/2019/05/14/uks-top-diplomat-urges-decisive-defense-budget-boost/. Accessed 7 Jan 2022
6. Clausen C (2017) Drones targeting providing freedom from terror: RPAs help reclaim Sirte. Air Combat Command, 432nd Wing/432nd Air Expeditionary Wing Public Affairs; August 1, 2017. https://www.acc.af.mil/News/Article-Display/Article/1265247/providing-freedom-from-terror-rpas-help-reclaim-sirte/. Accessed 7 Jan 2022
7. Congressional Budget Office (2017) The United States military force structure: a primer. Chapter 3. https://www.cbo.gov/sites/default/files/114th-congress-2015-2016/reports/51535-fsprimerbreakoutchapter3.pdf. Accessed 7 Jan 2022
8. Cook P (2016) Statement by Pentagon Press Secretary Peter Cook on US Air Strike in Libya. US Department of Defense, Washington DC. https://www.defense.gov/News/Releases/Release/Article/881794/statement-by-pentagon-press-secretary-peter-cook-on-us-air-strike-in-libya/source/GovDelivery/. Accessed 7 Jan 2022
9. Crawford N (2019) United States budgetary costs and obligations of post-9/11 wars through FY2020: $6.4 trillion. Costs of war research series. Watson Institute of International Affairs, Brown University. https://watson.brown.edu/costsofwar/files/cow/imce/papers/2019/US%20Budgetary%20Costs%20of%20Wars%20November%202019.pdf. Accessed 3 March 2021
10. Daragahi B (2018) ISIS resurrection: Libya attacks foreshadow terror to come. The Daily Beast; May 29, 2018. https://www.thedailybeast.com/isis-resurrection-libya-attacks-foreshadow-terror-to-come. Accessed 7 Jan 2022
11. Drew J (2015) US "carrier gap" could see naval power dip in the Gulf region. Washington DC; July 8, 2015. https://www.flightglobal.com/news/articles/us-carrier-gap-could-see-naval-air-power-dip-in-gulf-414440/. Accessed 7 Jan 2022
12. Eckstein M (2016) Amphib, Destroyer pairing seen in high-end exercise, current 6th fleet operations. US Naval Institute. https://news.usni.org/2016/09/20/amphib-ddg-pairing-seen-in-high-end-exercise-current-6th-fleet-operations. Accessed 7 Jan 2022
13. Evans G, Sahnoun M, et al (2001) The responsibility to protect: report of the international commission on intervention and state sovereignty. International Commission on Intervention and State Sovereignty, Ottawa. https://www.idrc.ca/en/book/responsibility-protect-report-international-commission-intervention-and-state-sovereignty. Accessed 7 Jan 2022

14. Filipov D, Roth A (2016) Russian jets keep crashing, and it may be an aircraft carrier's fault. The Washington Post; December 5, 2016. https://www.washingtonpost.com/world/russia-just-lost-its-second-warplane-in-a-month-of-syria/2016/12/05/f8353be8-bad8-11e6-817f-e3b588251d1e_story.html. Accessed 7 Jan 2022
15. Galbraith J (1983) The anatomy of power. Houghton Mifflin Company, Boston
16. Ibrahim A (2020) Rise and fall? The rise and fall of ISIS in Libya. Monographs, Books, and Publications. 913. https://press.armywarcollege.edu/monographs/913. Accessed 17 Nov 2020
17. Jasko K, Kurglanski A, Bin Hassan ASR, Gunaranta R (2018) ISIS: its history, ideology, and psychology. In: Woodward M, Lukens-Bull R (eds) Handbook of contemporary Islam and Muslim lives. Springer, Cham. https://d1wqtxts1xzle7.cloudfront.net/56391200/ISIS-chapter.pdf?1524451953=&response-content-disposition=inline%3B+filename%3DISIS_Its_History_Ideology_and_Psychology.pdf&Expires=1605277024&Signature=KiQjLlKukBVSvrob-AjdKD1JJ6~y5or0FgzMjQRwwmmS9rkqpkdWfa0sITEafmNrrDZISFu0unMP5Dmo5dBLhDH00Bx5Z6XVEjFLj9aJWlZK4XZlfYcoAzcc~hnGES7W-Uivt40nFNxVydXht6IOm~xKsXQYjldV5mUxR0~ubvKZBF01aLLX9T7H-8iaxj5B66wtOKz4gt7IoYzhnrklN87WrJvLNzF2ie4kl53WldYtU2iUwZWtcrZZNC0YOn69c5TSE8S9Cg1yKzgpnWcmcOdO0lN3Ti-o97iiwkuxbgLSItZzKD0VbM1Oe7RQ1GReR0BZhxXn0gW8AF03IaD~5A__&Key-Pair-Id=APKAJLOHF5GGSLRBV4ZA. Accessed 17 Nov 2020
18. Jervis R (2017) Perception and misperception in international politics. Princeton University Press, Princeton
19. Karasapan O (2017) The tragedy of migrants in Libya. Brookings Institute. May 15, 2017. https://www.brookings.edu/articles/the-tragedy-of-migrants-in-libya/. Accessed 25 Nov 2024
20. Kipling R (1919) We and they. The Kipling society. https://www.kiplingsociety.co.uk/poem/poems_wethey.htm. Accessed 25 Nov 2024
21. Marine Corps Air Station New River (2019) MV-22 Osprey. https://www.newriver.marines.mil/About/Aircraft/MV-22/. Accessed 7 Jan 2022
22. McGraw W (2016) Email correspondence dtd Thursday, December 8, 2016, 2:51 PM
23. Moltke H, Speri A, Turse N (2018) Secret war. The Intercept; June 20, 2018. https://theintercept.com/2018/06/20/libya-us-drone-strikes/. Accessed 7 Jan 2022
24. Morgan M (2017) Producing new spaces of violence: Libya and the changing style of NATO interventions after the global financial crisis. Geopolitics 22(4):911–933. https://doi.org/10.1080/14650045.2017.1293044. Accessed 7 Jan 2022
25. National Transition Council (2019) http://ntclibya.org/. Accessed 7 Jan 2022
26. Pack J (2022) Libya and the global enduring disorder. Oxford University Press, New York
27. Parmar I (2014) Obama, wikileaks, and American power. In: Obama and the world: new directions in US foreign policy. Routledge, pp 243–257. https://openaccess.city.ac.uk/id/eprint/16934/. Accessed 22 June 2021
28. Revkin M (2016) ISIS' social contract. Foreign Affairs. Retrieved from https://www.foreignaffairs.com/articles/syria/2016-01-10/isis-social-contract. Accessed 17 Nov 2020
29. Schogol J (2016) Amphibious assault ship Wasp, Marine Harriers exiting Libya operations. Marine Corps Times; October 21, 2016. https://www.marinecorpstimes.com/news/your-marine-corps/2016/10/21/amphibious-assault-ship-wasp-marine-harriers-exiting-libya-operations/. Accessed 7 Jan 2022
30. Snow S (2019) 22nd MEU Marines rocked ISIS fighters in Libya in 2016 with naval gunfire. Marine Corps Times; August 24, 2019. https://www.marinecorpstimes.com/flashpoints/2019/08/23/22-meu-rocked-isis-fighters-in-libya-with-naval-gunfire/. Accessed 7 Jan 2022
31. The Battle of Algiers (1966) Directed by Gillo Pontecorvo. Performances by Jean Martin, Saadi Yacef, Brahim Haggiag, and Tommaso Neri Written by Gillo Pontecorvo and Franco Solinas. Distributed by Rizzoli, Rialto Pictures
32. The White House, Office of the Press Secretary (2011) Text of a Letter from the President to the Speaker of the House of Representatives and the President Pro Tempore of the Senate. March 21, 2011. https://obamawhitehouse.archives.gov/the-press-office/2011/03/21/letter-president-regarding-commencement-operations-libya. Accessed 7 Jan 2022

33. United Nations Security Council (2011) Resolution 1973. https://www.nato.int/nato_static/assets/pdf/pdf_2011_03/20110927_110311-UNSCR-1973.pdf. Accessed 7 Jan 2022
34. United Nations Security Council (2015) Resolution 2259. Libyan Political Agreement. http://unscr.com/en/resolutions/doc/2259. Accessed 7 Jan 2022
35. Unknown author (2011) Qatar, several EU states up for Libya action: Diplomat. EU Business. https://www.eubusiness.com/news-eu/libya-unrest-summit.95v/. Accessed 7 Jan 2022
36. USS Wasp ARG Public Affairs (2016) Wasp amphibious ready group completes deployment. Navy News Service; December 24, 2016. https://www.navy.mil/submit/display.asp?story_id=98289
37. Vandiver J (2016) USS San Antonio joins Libya operation. Stars and Stripes, October 21, 2016. https://www.stripes.com/theaters/uss-san-antonio-joins-libya-operation-1.435175#:~:text=STUTTGART%2C%20Germany%20%E2%80%94%20The%20Navy%E2%80%99s%20USS%20San%20Antonio,planned%20switch%2C%20U.S.%20Africa%20Command%20said%20on%20Friday. Accessed 24 March 2021
38. Vucetic S (2011) Anglosphere: a genealogy of racialized identity in international relations. Stanford University Press, Redwood City
39. Watanabe L (2019) UN mediation in Libya: peace still a distant prospect. Center for Security Studies, Zurich. https://css.ethz.ch/content/dam/ethz/special-interest/gess/cis/center-for-securities-studies/pdfs/CSSAnalyse246-EN.pdf. Accessed 7 Jan 2022
40. Wrong D (1979) Power: its forms, bases, and uses. Transaction Publishers, London and New Brunswick

Chapter 4
The Annexation of Crimea by the Russian Federation and the Response of the United States

> The most revolutionary event of 1991 for Russia might not be the collapse of Communism, but the loss of something Russians of all political stripes think of as part of their own body politic, and near to the heart at that: Ukraine.
>
> United States Ambassador to Russia Robert Strauss, 1991

> When people begin to lean toward and rejoice in the reduced use of military force to resolve conflicts, war will be reborn in another form and in another arena, becoming an instrument of enormous power in the hands of all those who harbour intentions of controlling other countries or regions.
>
> Liang and Xiangsui [14]

This case study examines how Russia and the United States used their respective national power, in terms of the instruments and forms of power, to pursue their national interests during the 2014 Crimean affair in which Russia regained control of Crimea. Claiming that the right to self-determination supersedes the principle of territorial integrity based on precedents set by the United States and its allies in the late twentieth century, Russia justified its actions that spanned the four domains of national power, each in concert with the others, to facilitate the return of Crimea to Russian control, which has been a critical aspect of Russian national interest since the end of the seventeenth century [3, p. 77, 12, p. 111]. The United States—along with its allies in Europe, the United Nations, European Union, and the North Atlantic Treaty Organization, among others—rejected the validity of Russia's actions as a violation of international law and publicly condemned the action as an unlawful invasion of sovereign Ukrainian territory followed by economic sanctions against Russia and increased military investment in Ukraine. In this chapter, beginning with a brief review of the salient historical context characterizing the relationships between the immediate parties to the conflict—Russia, Ukraine, and Crimea—facilitates subsequent analysis of how both Russia and the United States exercised their national power and why each state justified its actions as part of its national interest. The Russian invasion of Ukraine in February 2022 falls beyond the scope of this case

K. L. Wedekind, *Constructing Power: A Critical and Constructivist Theory Using the American Case*, Advanced Sciences and Technologies for Security Applications,
https://doi.org/10.1007/978-3-031-84893-3_4

study though the findings in this chapter can serve as a prelude for subsequent analysis of the 2022 conflict that at first glance will hold up under scrutiny.

The Crimean episode demonstrates how national identity and interest relate to the sources, instruments, and forms of power with juxtapositions between intentionality and effectiveness shown by the Russian Federation and the United States. Russia acts with a high degree of intentionality across time measured in centuries and familiar space bound by common history, language, and culture while also working across mutually reinforcing lines of effort to increase effectiveness by connecting complementary sources, instruments, and forms of power. The Russian approach set conditions for its exercise of condign power through the use of force through information operations across decades supported by public diplomacy bilaterally with Ukraine and multilaterally through international organizations such that when Russia occupied and annexed Crimea the international community could do little but accept it as the realization of a self-fulfilling prophecy; Russia exercised its power to satisfy its interest by recovering the Crimean peninsula under the pretext that doing so reunited a dissonant aspect of Russian national identity, made all the more significant by the access to the global commons that Crimea provides. The United States enters the fray too little, too late and exercises its power linearly—that is, along a single line of effort—by relying upon property, compensatory power, and authority to apply economic sanctions, supported only by the requisite public condemnation that generated the sanctions. The result is a decisive Russian victory not only over Ukraine but also the United States and the international community; while the tactical result returned Crimea to Russian control, the strategic implication of allowing the disruption of international norms to persist sets a new standard of acceptable behaviour within the community of nations—especially when considering the international response to comparable actions by Iraq against Kuwait hardly more than two decades prior; further analysis comparing the two incursions against Kuwait and Ukraine exceeds the scope of this book and must wait for subsequent works. At present it must suffice for this case study to contrast the actions of the Russian Federation and the United States leading up to and following Russian annexation of Crimea in 2014.

This chapter seeks to illuminate the role of identity and interest formation for Russia and the United States, as well as the consequent exercises of power by each and the final disposition of the episode. An examination of how Russia views itself links its cultural identity and national interest while providing insight into why and how Russia exercised its power. The Russian effort to satisfy its conception of national identity by annexing Crimea demonstrates the causal relationship between the identity, interest, and the exercise of power. The United States reaction also reflects its conception of national identity and interest, which determine the appropriate exercise of power in response.

It is critical to note that this case study does not address the national identity, interest, and power of Ukraine. The reason for this omission is the focus on compensatory and conditioned power in the form of retaliatory economic sanctions and official statements condemning the Russian annexation that the United States attempts to use against Russia to enforce norms within the international order that existed in

2014. The power theory in this work views power relations as transpiring between two and only two agents and uses the United States as the constant agent across all case studies, which makes the adversarial relation between the United States and Russia that manifests primarily in the economic and information domains in this episode appropriate for the case study.

It is possible to interpret the omission of Ukrainian perspectives as a denial of Ukrainian agency in this case study; this is not the case. Ukraine possesses its own national identity, interest, and power, and maintains power relations with the United States, Russia, and all other agents present in the international order with each respective relation a viable candidate for analysis with Ukraine as the focus and not the United States. The power theory offers a lens through which future research could examine Ukrainian power in relation with Russia, NATO, European Union, the United States, or other agents, especially since the beginning of Euromaidan in November 2013. An inquiry into how Ukraine exercised its power in 2014 compared to 2022 relative to these others could prove insightful and is ideal for future research based upon the present work.

4.1 In Search of the Southern Sea: Russian National Interest

Since the end of the 17th Century, Russia has sought to expand its southern border to Crimea so that it could maintain a significant naval presence in the Black Sea from which it could project power through the Mediterranean Sea to the world's oceans. Czar Peter I recognized that his ambitions to modernize Russia and increase its standing among the European powers would require giving it a naval capability from which Russia could project power on par with its Western rivals. Peter led a Russian expedition into Ottoman-held territory along the Sea of Azov and founded the city of Taganrog there, which served as the first base for the Russian navy and from which the Russian empire could challenge the Ottoman for control over the Black Sea and access to the Mediterranean [17, p. 147]. The foothold in the Sea of Azov and naval presence there proved ineffective, as Crimea and the Strait of Kerch remained under Ottoman control and limited Russian naval reach to the shallow Sea of Azov. Crimea would remain under control of the Ottoman Empire until the signing of the Treaty of Küçük Kaynarca in 1774 that ended the Russo–Turkish War and granted Crimea short-lived independence [10]. In the years that followed the Treaty, Empress Catherine II whittled away at Crimean independence by inciting rebellion against all pro-Ottoman leaders while at the same time sponsoring Russian settlement in Crimea.

In 1783, Empress Catherine II formally annexed Crimea into the Russian Empire. Crimea would remain part of the Russian Empire until its dissolution in 1917, and then fall under the control of the Union of Soviet Socialist Republics (USSR) in 1921, first as part of the Russian Soviet Republic until 1954 when it was transferred to the administrative control of the Ukrainian Soviet Socialist Republic (SSR), which itself was a part of the Union of Socialist Soviet Republics [28, p. 1]. Since the Union of Socialist Soviet Republics held legitimate authority over the Ukrainian SSR, and the Ukrainian SSR controlled Crimea, the administrative transfer of Crimea to the Ukrainian SSR proved superficial until 1992 when Ukraine gained its independence and Crimea became sovereign Ukrainian territory [19]. At that point, Russia lost Crimea and the city of Sevastopol to Ukraine, and with them unimpeded access to the Black Sea for the Russian Navy. While Russia and Ukraine did broker an agreement under the Partition Treaty on the Status and Conditions of the Black Sea Fleet of 1997 that divided the assets of the USSR Black Sea Fleet and leased to Russia the use of Sevastopol's naval facilities, the underlying and unresolved issue upon the signing of the Partition Treaty related to the true interest of both parties, which concerned the sovereignty of Crimea and control of Sevastopol:

> For Ukraine the issue was maintaining new-found independence from Russia. In seeking to reach a deal Ukraine wanted at all costs to avoid being bullied by Russia and to maintain Ukrainian sovereignty over Sevastopol and the rest of Crimea. Ukraine did not want to acquire an entire new "fleet" as it could not afford the maintenance costs for even a fraction of the ships. For Russia, the issue was in acquiring the ships and the rights to base them (preferably on sovereign Russian territory). Russia needed the remains of the [Black Sea Fleet] not for any strategic purpose but as a symbolic instrument to help it reassert power on its southern flank—vis-a-vis Turkey, the Caucasus, and future Caspian oil flows [5, p. 2].

In 1992, Russian and Ukrainian national interests in Crimea became irreconcilable. Ukraine needed to maintain its independence by preserving its hold on the three Westphalian principles of territorial integrity, managing its own domestic affairs without Russian influence, and to gain equal standing with Russia as a sovereign state. Russia needed to regain control of Crimea and Sevastopol, both at this point sovereign Ukrainian territory. The Partition Treaty in 1997 only delayed the conflict, made inevitable by an irresolvable conflict of interests and exacerbated by a resurgent Russia and a stagnant Ukraine in terms of national power.

4.2 The Long Road to War: Reconstructing Russian Identity

On April 3, 2008, NATO declared in the Bucharest Summit Declaration that "NATO welcomes Ukraine's and Georgia's Euro-Atlantic aspirations for membership in NATO. We agreed today that these countries will become members of NATO" [20]. The inevitable addition of the Republic of Georgia and Ukraine to NATO compelled Russia to exercise its power to prevent, or at least delay, the extension of NATO Article 5, which requires all NATO members to consider an attack against one as an attack against all, to apply to states upon which Russian national interest depends. Having both Ukraine and Georgia incorporated into NATO would hinder Russia's ability to regain several disputed territories; from Ukraine, Crimea and Sevastopol, and from Georgia, Abkhazia and South Ossetia, two provinces that had attempted to secede from Georgia and join Russia since 1991 and 1993, respectively. For Georgia to begin the process to join NATO, it would need to hold its parliamentary elections in May and "achieve acceptable international standards, and that it must make progress on resolving its two "frozen" conflicts within its territory;" Ukraine would need to convince the majority of its population that opposed NATO membership of the value to be gained from closer ties to Europe and the United States, as opposed to Russia [7, p. 5]. The Bucharest Summit Declaration made Russian action in both Ukraine and Georgia inevitable, and necessary from the Russian perspective to preserve its national security interest. Russia capitalized on Georgia's attempt to reintegrate Abkhazia and South Ossetia in August 2008 by devastating Georgian forces and moving its own to reinforce rebels in the two provinces, thereby perpetuating the conflicts and thus precluding Georgia's advancement in the membership action plan (MAP) process.

Russia would not receive such a comparable opportunity from Ukraine. Rather, Russia increased its efforts within the information domain, through which it had been working since at least 2003 when the Russian President stated that it is "completely unacceptable that the cradle of the Russian state, the mother of the Russian cities, Kiev, also the site of the ancient Kiev Pechersk Lavra, lies beyond the boundaries of the current Russian Federation" [11, p. 3]. The motivation indicated in this statement identifies a schism in Russian national identity created by an independent, sovereign Ukraine; it also implies that the Russian Federation did not view its borders at the time as fixed in time and space, but rather temporary and apportioned. Russia views its national identity as constituted by shared history and values that transcend the boundaries of sovereign states as opposed to confined by the strictures of borders that correlate legitimate authority and territory. Perceiving the legitimate authority ascribed to sovereign states as subordinate to shared history and culture liberates the Russian Federation from the international norms that discourage encroachment by one state upon another in cases where common national identity exists. Invoking shared identity as justification for territorial expansion bucks the norm when it transpires through force as opposed to authority; this suggests that states use force to reshape international norms that distinguish between the several forms of authority. Force provides a way for states to modify the limits of acceptable behaviour, though each use of force entails reciprocal action that the state must endure for its exercise of power to retain its properties of intentionality and effectiveness. Utilizing conditioned power to reshape expectations of state behaviour reinforces the exercise of condign power and depends upon convincing others to accept the outcome. The idea that Russia held a legitimate claim to parts of Ukraine would serve as the primary message used by Russian information operations to persuade, in an act of conditioned power, two primary audiences—the people of Ukraine and the United States—in the validity of a Russian-controlled Crimea. Russia would work to persuade enough of the Ukrainian citizenry that its true national identity is Russian and thus belonged within the Russian Federation so that the impending incursion would face less intense opposition. Russia would need to justify its actions to the United States, and by extension the rest of the West, by arguing that the principle of self-determination supersedes that of territorial integrity, and that Ukraine, or at least parts of it, truly had always belonged to Russia.

This message resonates in several key remarks made by the Russian President between 2008 and 2014. During a private meeting that took place at the NATO meeting in Bucharest, the Russian President explained the situation to the President of the United States: "You don't understand, George, that Ukraine is not even a state. What is Ukraine? Part of its territories is Eastern Europe, but the greater part is a gift from us" [16]. This outright rejection of Ukrainian sovereignty reverberates throughout speeches delivered by the Russian President between 2008 and 2014. The Russian President alluded to such ambitions in July 2013 in Kiev at the 1025th anniversary of the baptism of Rus: "This is not just Ukraine's civilisational choice…at [this] baptismal site on the Dnieper River, a choice was made for the whole of Holy Rus, for all of us. Our ancestors who lived in these lands made this choice for our entire people…we know that…there are the Ukrainian people and the Belarusian people, and other peoples too, and we respect all the parts of this heritage, but at the same time, at the foundations of this heritage are the common spiritual values that make us a single people" [29]. Here the Russian President projects the idea of common heritage that unites the Rus peoples beyond state borders into something described as the Russian World. A critical linguistic shift in this speech betrays its true intentions by "using with the name Ukraine preposition 'in' that in the Russian language is used with the state, but not territory, and that Mr. Putin was using before regularly, to preposition 'on' that in the Russian language is used with a territory, not a state. So, instead of saying 'in Ukraine' he started to say 'on Ukraine' with a very distinct flavour of non-recognition of independence and sovereignty of the Ukrainian state" [11, p. 4]. The so-called "choice" is rather an ultimatum akin to that set forth by Benjamin Franklin to the American colonies prior to their revolution—"to join, or die" [6]. To convince the audience, the Russian President subtly denies Ukraine's sovereignty and then uses third party amplification to support his position by invoking the acquiescence of the Ukrainian Orthodox Church based on the anticipation that public opinion of Russians can be influenced by the personal authority held by the clergy. This speech, and others like it, is an act of conditioned power intended to persuade the audience that the borders separating Russia and Ukraine exist as little more than lines on the map; such thinking implicitly undermines Ukrainian sovereignty in favour of the union of old.

Russia's interest in Crimea proves singular. It seeks to regain the legitimate authority it once held over Crimea to reclaim access to the global commons unencumbered by the limits stipulated in the Partition Treaty, or any other agreement, so that it might once again project naval power across the world's oceans. The ability to project naval power globally would allow Russia to exercise greater influence over global affairs and contribute to the restoration of Russia's former standing in the domestic, regional, and international order. The ideology contained in the Russian World concept and the corresponding information operations intended to condition through persuasion provides one aspect of the means to the end in which Russia controls Crimea. To do so, Russia undermined Ukrainian sovereignty by setting conditions that would facilitate diplomatic and military actions against Ukraine that once complete would leave Crimea and Sevastopol under Russian control. By exercising conditioned power intended to influence the beliefs held by parties to the inevitable conflict through an information campaign with a singular message, Russia set conditions for its success in 2014, years before it initiated operations in the other three domains of national power.

With political conditions sufficiently deteriorated, Russia exploited the opportunity to secure its most valuable former territory in Crimea and Sevastopol produced by its shaping actions in the information domain. The second phase would rely upon the diplomatic and military domains of national power, as opposed to the first which Russia conducted primarily through the information domain. The catalysing event occurred in February 2014 when the Ukrainian Parliament impeached then-President Viktor Yanukovich in response to his refusal to sign the Deep and Comprehensive Free Trade Agreement with the European Union in favour of a combined loan and gas agreement with Russia; pro-European protests erupted forcing Yanukovich to abandon his post. Meanwhile, "Russia perceived the power shift as a coup d'état and responded by annexing the Crimean Peninsula in March 2014... and by launching an undeclared 'hybrid war'—involving strong economic pressure and targeted disinformation measures—against Ukraine. The EU, the US and a number of other countries imposed sanctions against Russia, which retaliated with countersanctions" [1, p. 2]. Russia's actions, the reaction from the United States and its allies, and the Russian counteraction constitute exercises of national power with respect to the interest of each party to the conflict, which traces its origins to the national interest of each state; that is, Russian national identity informs Russian national interest and the consequent exercise of Russian national power just as United States national identity informs its national interest and exercise of national power. The critical difference between how Russia and the United States exercised power within the Crimean episode arises from a disparity in comprehensiveness, intensiveness, and extensiveness that Russia achieved by integrating multiple power domains in a mutually reinforcing way, as opposed to the United States and its relatively limited reaction. Within this chapter, Fig. 4.1 illustrates the Russian exercise of power while Fig. 4.2 shows the United States response; the contrast between the two explains the enduring success of the Russian endeavour.

4.3 Exercising Mutually Reinforcing Domains of National Power

Russia integrated the three additional domains of national power into its initial actions against Ukraine in March 2014. Information operations, having begun a decade prior, would continue and remain Russia's main effort but reinforced with supporting diplomatic, economic, and military operations intended to assure the overall success of its endeavour. These supporting activities would prove relatively limited in duration and applied to specific purposes that would achieve tangible effects, rendering the status

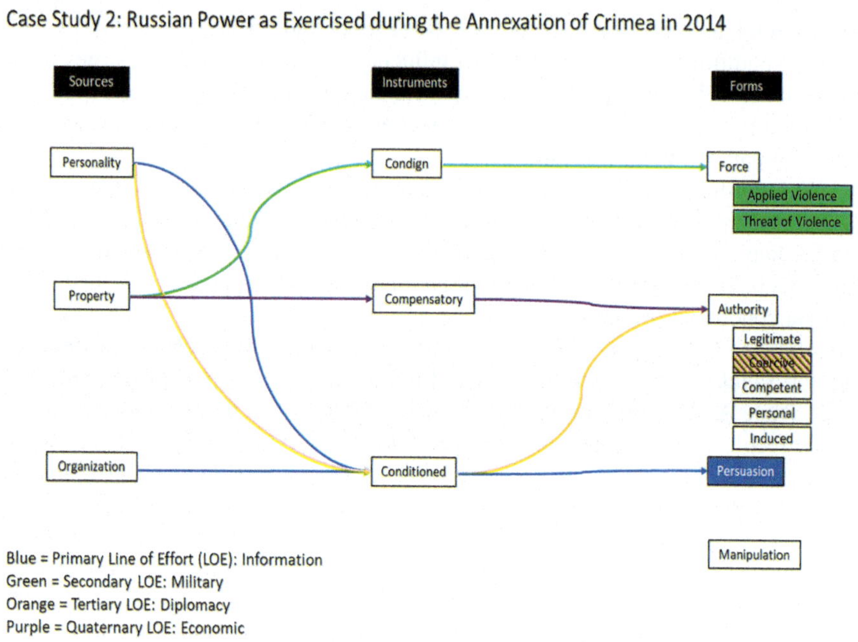

Fig. 4.1 Russian power as exercised during the annexation of Crimea in 2014 (Author's own)

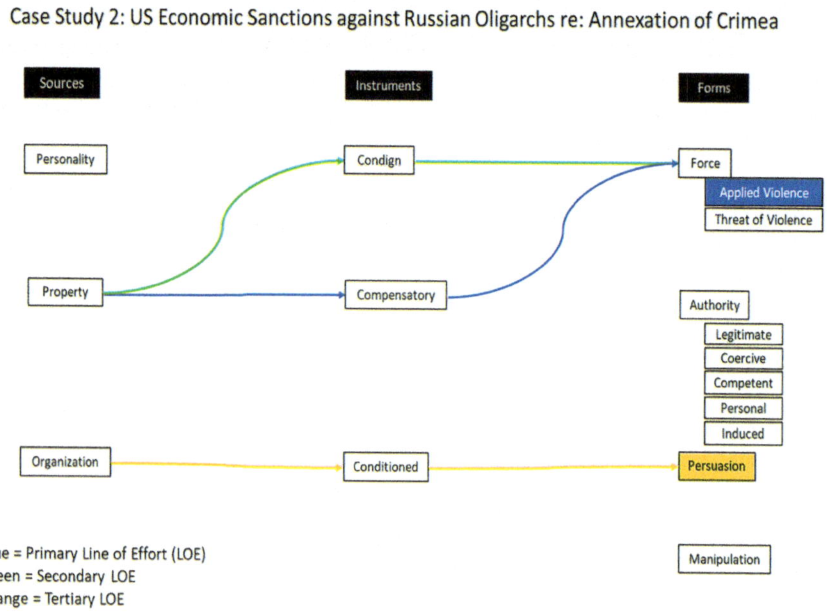

Fig. 4.2 US exercise of power against Russia in response to its annexation of Crimea (Author's own)

quo obsolete. Diplomatic actions would directly support the purpose of the information operations, while military and economic actions would support the diplomatic directly and ultimately by extension the information operations comprising Russia's main effort.

This figure uses process tracing to outline in detail the Russian exercise of power in annexing Crimea in 2014. All sources, instruments, and forms of power except manipulation appear in combination with one another along complementary lines of effort that reflect the national power domains. The comprehensiveness of power exercised manifests as effectiveness in the Russian operation and its success. The model shows how Russia employed all sources, instruments, and forms of power to achieve its goal of annexing and holding Crimea. Four mutually reinforcing lines of effort reflecting the domains of national power operate within and across the sources, instruments, and forms upon which this work focuses as the substrate of power itself and national power specifically.

Russia's exercise of conditioned power through the form of persuasion transpires within the information domain and constitutes its main effort because Russia must convince Ukraine and the rest of the world that Crimea actually belongs under Russian and not Ukrainian control. Conditioning its opposition to alter its beliefs is the critical element of the operation. Russia could conduct an overt military operation to invade Ukraine and retake Crimea, but such an action incurs an increased risk of an international military response to restore the status quo ante without setting conditions for acceptability by modifying the beliefs of others, specifically those interested but uninvolved parties like the United States unilaterally as an individual and NATO multilaterally as a collective. By working through the information domain, supported directly by diplomacy backed by hybrid warfare and economic pressure dilutes the overall Russian operation to something less than the threshold beyond which the international community would be compelled to conduct a major military operation. It becomes easier to accept Russia's action if the potential opposition doubts its conviction in any potential reaction—this is especially so when it comes to multilateral military action. While overt military action nearly guaranteed immediate success, the long-term prospect of being rebuffed would leave Russia without Crimea and access to the Black Sea. By prioritizing its desired end state above all else, Russia opted to avoid a major military confrontation to increase its chances of retaking and holding Crimea by orchestrating its action across the four domains of national power.

The twenty-four days between the then-Ukrainian President's abandonment of his position on February 22 and Russia's annexation of Crimea on March 18 proved the most critical period for the entire, decade-long Russian operation. Russia's actions during this time spanned the full spectrum of national power domains, as it brought to bear diplomatic, military, and economic power in support of its ceaseless work in the information domain. The immediate purpose of this phase of the Russian operation was to seize Crimea; to do so, Russian information operations worked through three lines of effort concurrently to "[discredit] the new government in Ukraine, [emphasize] the grave danger to Russians in Ukraine, and [ensure] the display of broad support for Crimea's 'return home' to the safety of Russia" [13,

p. 13]. These information operations, which are the manifestation of conditioned power through a combination of persuasion and manipulation, would continue to shape perceptions in Ukraine and the West with the former directly targeted and the latter observing as an interested third party.

On February 26, the day before military forces occupied the Crimean Parliamentary building, key Russian political influencers—among them the leader of Spravedlivaya Rossiya and the Chechen Republic—delivered messages on Russian television intended to appeal to one of the greatest fears common among ethnic Russians living in Crimea and greater Ukraine by declaring the Ukrainian regime change and interim government illegitimate, born of nationalism and embracing fascist ideology, and with the ultimate goal of assimilating ethnic Russians into a new Ukraine in possession of its own identity and culture. According to this messaging, the first step for the interim Ukrainian government would be to "force Russians in Ukraine to abandon the Russian language," supposedly intended to justify and necessitate Russian intervention into Ukrainian affairs under the guise of protecting ethnic Russians living in Crimea and Ukraine [13, p. 14]. While Russian politicians presented the message from positions of legitimate authority, an alleged grassroots organization, Stop Maidan, began propagating the same message among ethnic Russian communities in Crimea using visual advertisements such as tents emblazoned with logos, direct telephonic calling, and quips posted on signs and banners implying that the regime change in Ukraine had been foreign-inspired by fascist extremists emanating from Europe; individuals working under the auspices of the Stop Maidan group went so far as to collect signatures on petitions calling for greater autonomy for Crimea while also attempting to condemn pro-Maidan individuals as being controlled by the West, specifically the United States, and working against what could be considered the Crimean interest [13, p. 15]. These efforts constitute exercises of national power within the information domain targeting audiences in Ukraine and Russia to influence their willingness to accept the imminent annexation of Crimea; it would fall to the Russian President to set a similar tone to all third parties to the conflict within the international community.

4.4 Messaging as Conditioned Power and Persuasion

During a press conference held on March 4, the Russian President presented the case for intervention and the eventual annexation of Crimea by the Russian Federation while at the same time denying that Russia would take such actions or had any involvement in the set of circumstances then unfolding in Ukraine and Crimea. Maintaining some semblance of plausible deniability remained critical to the success of the Russian endeavour at this point, as Crimea still fell under Ukraine's legitimate authority and little had been accomplished that could sway international opinion otherwise [13, p. 24]. The necessity of justifying an exercise of power while at the same time denying that it has occurred compels the revelation of details even through simply acknowledging the possibility of the actions; that is in this instance, presenting

Russia's case for intervention while simultaneously denying its possession of any intentions to annex Crimea or activities undertaken to that end betrays the motive and the means through which Russia intended to accomplish its objective.

Equivocations in the Russian President's remarks on March 4, 2014 make veiled references to the ulterior motives and covert actions being undertaken by Russia in Ukraine at that time. Closer examination indicates that each of these statements contains elements of rhetorical fallacies that, though intended to introduce doubt as to the true purposes of the Russian action, correlate to events unfolding in real time unbeknownst to all except the Russian national security apparatus. Use of ambiguity in official remarks is unsurprising and commonly expected from heads of state; such practices prevent the premature disclosure of state secrets, to include those concerning national interest and the exercise of national power. The Russian President's press conferences from March 4 prove noteworthy because the construction of the argument presented correlates to Russian actions that remained secret at the time; that is, hindsight has shown that the denial of Russian exercises of power in Ukraine proved false, thereby suggesting that these remarks constitute not only an exercise of conditioned power, but an exercise of manipulation intended to deceive all who would receive its message.

To begin, the Russian President introduces a false dichotomy to limit the number of plausible explanations only to one that complies with the manufactured narrative surrounding the turn of events in Ukraine: "There can only be one assessment: this was an anti-constitutional takeover, an armed seizure of power. Does anyone question this? Nobody does. There is a question here that neither I, nor my colleagues, with whom I have been discussing the situation in Ukraine a great deal over these past days, as you know—none of us can answer. The question is why was this done?" [31]. This statement characterizes the destabilization of Ukraine as an illegal action with an unknowable purpose performed not by Russia, but some other phantom organization standing opposed to the Ukrainian constitution.

A retrospective deterministic claim follows, which argues that the instability in Ukraine became inevitable upon its independence from Russia: "This revolutionary situation has been brewing for a long time, since the first days of Ukraine's independence" [31]. This statement implies that the "revolution" in Ukraine should be viewed as not only inevitable, but a natural process of reconstituting a fractured state, which from this perspective would be Russia on account of its loss of the parts of Ukraine where ethnic Russians comprise a majority of the population. The Russian President then justifies the previous premise through the fallacy of composition and division: "Most importantly, people should have the right to determine their own future, that of their families and of their region, and to have equal participation in it. I would like to stress this: wherever a person lives, whatever part of the country, he or she should have the right to equal participation in determining the future of the country" [31]. This comment blends the egalitarian Westphalian principle with the contemporary notion of self-determination but applies it too acutely to be valid within the international order that views the state—not regions, peoples, or families independent from one another, but taken as a single political unit bound in geographic space. Self-determination and equality remain valid principles, but only in possession of the state and not at any level beneath it, which causes the property of emergence at the state level and none other [9].

What follows is a refutation of a presumptive correlation between Russian military exercises and the actual military operations in Ukraine through an appeal to consequences: "Regarding the deployment of troops, the use of armed forces. So far, there is no need for it, but the possibility remains. I would like to say here that the military exercises we recently held had nothing to do with the events in Ukraine. This was pre-planned, but we did not disclose these plans, naturally, because this was a snap inspection of the forces' combat readiness" [31]. Associating training exercises with the deployment of combat troops into a sovereign state—also known as an invasion—would lend credence to speculation that Russia had engaged in a military operation in Ukraine. The complexity of the entire operation suggested that Russia had planned long enough in advance to coordinate the deployment of military units with complementary capabilities which were required to execute a complex covert operation. The information operations created a diversion and constituted an exercise of conditioned power through the strategic application of persuasion and manipulation. The equivocation in this statement, and throughout the March 4 press conference, maintains Russia's deniability while introducing doubt into the beliefs of the audience concerning the true nature of the situation then-unfolding in Ukraine and what the appropriate response should be if the ideas espoused by Russian information operations were at all truthful. Doubt in this case appears as the effects of information as a weapon targeting third parties to the conflict and intended to lessen their collective resolve to react with condign power, specifically the use of physical force to project power.

The Russian President continues by introducing an even-if fallacy based on false premise and attribution by co-opting the audience by projecting knowledge upon them that they would otherwise have no way of possessing: "As you may know, we have a direct appeal from the incumbent and, as I said, legitimate President of Ukraine, Mr. Yanukovych, asking us to use the Armed Forces to protect the lives, freedom and health of the citizens of Ukraine" [31]. The purpose of this statement is to impress upon the audience that even if Russia were to use its military that action would be legitimate as it would be undertaken at the request of the Ukrainian President who had abdicated his post as such.

Perhaps the most consistent aspect of Russian information operations has been the message's focus on the commonalties between Ukraine and Russia that supposedly surpass state borders, rendering such divisions as little more than lines of a map. This argument exemplifies the association fallacy, wherein the properties of one entity become projected upon another and taken as such despite contradictory evidence. The Russian President's description of Ukraine in 2008 as less than a sovereign state becomes watered down in the March 4 press conference: "We have always considered Ukraine not only a neighbour, but also a brotherly neighbouring republic, and will continue to do so" [31]. Describing Ukraine in a familial sense affords Russia with the ability to assert itself as the father of the prodigal son rather than the older, stronger, and angrier brother, which it more closely resembles. The purpose of this language is to replace the egalitarian relationship that sovereign states maintain with one another with another kind of relationship altogether, even if only in the perception of the interested third parties.

The next fallacy presented by the Russian President is an appeal to ridicule intended to undermine the legitimate authority of the interim Ukrainian government: "If I were in the shoes of those who consider themselves the legitimate authorities, I would not waste time and go through all the necessary procedures, because they do not have a national mandate to conduct the domestic, foreign and economic policy of Ukraine, and especially to determine its future" [31]. The remarks here deny the Ukrainian government's right to rule, including the exercise of legitimate authority and by extension the exercise of national power for any purpose such as resisting the inevitable Russian annexation of Crimea.

The penultimate fallacy is one of relevance. The Russian President cites the precedent set by the United States in its actions while fighting the Global War on Terrorism:

> We are often told our actions are illegitimate, but when I ask, 'Do you think everything you do is legitimate?' they say 'yes.' Then, I have to recall the actions of the United States in Afghanistan, Iraq and Libya, where they either acted without any UN sanctions or completely distorted the content of such resolutions, as was the case with Libya. There, as you may know, the resolution only spoke of closing the airspace for government aircraft, while it all ended with bomb attacks and special forces land operations...Our partners, especially in the United States, always clearly formulate their own geopolitical and state interests and follow them with persistence. Then, using the principle 'You're either with us or against us' they draw the whole world in. And those who do not join in get 'beaten' until they do [31].

The crux of this argument hinges on the question of legitimate interventions into sovereign states. By citing the United States' exercise of power addressed in the previous chapter on Libya among others, the Russian position cites the actions of the United States without consideration of the motivation in favour of the consequence, whereas the position of the United States depends upon the reasons for intervening in the affairs of sovereign states. The conflicting positions represent competing perspective concerning jus ad bello, or what can be considered legitimate reasons for going to war without actually declaring war. The condition wherein states do not declare a state of war but engage in warfare across all domains of national power represents one of the key aspects of this work that proves most interesting; while the sources, instruments, and forms of power prove finite in theory and limited in reality, the ways in which an agent like the state may combine them prove infinite and reflect the uniqueness of the agents, their engagement, and the circumstances characterizing their interaction. The theory herein offers a heuristic framework that can account for the multitude and diversity of factors influencing the way in which each engagement transpires and its consequent outcome.

The climax of the Russian President's argument unites each fallacy under a single, overarching modal fallacy intended to validate the imminent Russian annexation of Crimea and the operations spanning the gamut of national power domains being executed at that time in and against Ukraine: "If I do decide to use the Armed Forces, this will be a legitimate decision in full compliance with both general norms of international law, since we have the appeal of the legitimate President, and with our commitments, which in this case coincide with our interests to protect the people with whom we have close historical, cultural and economic ties. Protecting these people is in our national interests" [31]. In fact, the Russian military had been conducting operations in Ukraine for several days with the decision to use the military had been made far enough in advance to allow for task organization, training, and rehearsals of concept that would be required to execute a complex, covert operation that relied not only on the use of the violent component of condign power, but also the conditioning effect of simply placing personnel mimicking the ways in which uniformed military forces would conduct themselves while seizing and occupying foreign territory. Russia's case is that its actions, represented as hypothetical in these remarks but transpiring concurrently with their delivery, would prove justified on account of a request from the then former Ukrainian President to do so and the historic ties that bind the fate of Russia and Ukraine—when in fact neither reason taken separately or together could, or would, legitimize the annexation of Crimea.

The remarks delivered by the Russian President during the March 4 press conference serve as a pre-emptive defence of Russia's actions and indicate that it exercised its national power to manufacture the crisis in Ukraine to justify Russia's invocation of the United Nations Responsibility to Protect doctrine to justify its annexation of Crimea. Russian information operations targeted what it perceived as existing insecurities in the various population groups party to the conflict over Crimea to influence the perceptions of different groups, whether through affirmation and amplification or by sowing the seeds of distrust and discord. While Russia utilized information—drawing upon its national identity, or personality, as a source of power exercised through multiple acts of persuasive conditioned power in the form of Presidential remarks with a consistent, sustained message—as the primary mechanism to execute its operation, it could not have achieved its desired end state without the exercise of diplomacy, use of military force, and exertion of economic pressure to support its information campaign, which relied upon the use of conditioned power in the form of persuasion and manipulation. These activities set conditions that made possible the supporting diplomatic, military, and economic exercises of power through their respective instruments and forms. The national identity that Russia defined for itself—or rather, the Russian role-taken—included Ukraine and specifically Crimea as part of its territory while Ukraine and all other states recognized Crimea as integral to Ukraine. This disparity in identity led to a difference in national interest between Russia and Ukraine and led to the Russian exercise of power to annex Crimea.

4.5 Diplomacy in Action

One week after the Russian President's remarks of March 4, the Crimean Parliament stated its intent to declare independence from Ukraine and seek immediate admission to the Russian Federation pending the results of a public referendum held on March 16. The referendum results supported the declaration of independence that had been signed on March 11. Crimea and the city of Sevastopol were admitted to the Russian Federation on March 18 when the Russian President signed two laws—one admitting Crimea and declaring Sevastopol a city of federal importance and the other ratifying the terms of agreement between Crimea and Russia [32].

The diplomatic actions taken first by Crimea to declare independence, hold a referendum vote to confirm its declaration, and then to seek admission to the Russian Federation, and followed by Russian acceptance of Crimea and granting special status to Sevastopol were made possible through the application of national power in the form of information intended to condition the perception and beliefs of the peoples of Crimea, Ukraine, and Russia to see this transfer of authority as a natural and legitimate act. Having succeeded in cultivating this belief, Russia made these diplomatic efforts a fait accompli. With the peoples directly impacted by the transfer of Crimea to Russia prepared to accept its inevitability, Russia was able to reduce staunch opposition to its operation and while resistance did occur, and still exists, it did not prove sufficient to garner a supporting international military response. The Crimean declaration of independence, referendum, and appeal to the Russian Federation as well as Russia's acceptance account for the diplomatic actions that were necessary to complete Russia's annexation of Crimea while remaining below the threshold of justification for a third-party military response and appearing as the fulfilment of the underlying goals of a decade's worth of Russian information operations.

4.6 Polite People and Little Green Men in Crimea

Whereas the Russian diplomacy worked to legitimize its actions, Russian military operations were conducted in such a way that Russia could retain plausible deniability, and the ability to retrench or disengage if necessary, about the use of military force to execute the annexation and thereby protect the credibility of its information operation and the overall endeavour. "Russia practiced tactical, operational, and strategic deception, deceiving the local population, the Ukrainian leadership, and the West" through the use of *maskirovka*, which "is an ancient operational art of concealment and camouflage in the conduct of conventional operations" [13, p. 23, 27]. Russian soldiers manipulated locals into believing that the Russian soldiers were Ukrainian self-defence forces or police units and as such posed no threat to the local Ukrainian population while public messaging at the national level "confused Ukraine's leadership as to Russia's true intentions and thereby delayed a Ukrainian response" [13, p. 23]. The international community, notably the United States and its allies, "mistook Russian tactics aimed to create plausible deniability as signs of an effort to negotiate a political settlement and then de-escalate, rather than annex the peninsula. Western officials urged caution and sought to freeze conflict, thinking that Moscow might be seeking an off-ramp to the crisis. The deception also afforded Russia multiple points for disengagement and plausible deniability should the operation have gone awry. It was, in effect, Russia's exit strategy" [13, p. 23]. By design,

Russia integrated its operations across the domains of national power by reinforcing each line of effort with the others—information set conditions that facilitated diplomatic manoeuvring that disguised military operations, which Russia would hold using the threat of economic power [27]. Conditioned power in the form of information operations and diplomatic posturing exercised through persuasion and manipulation facilitated condign power in the form of force through applied violence; the threat of the use of economic force after the fact make the cost of action seeking to revert the annexation too great for both the immediate adversary and any third parties. Integrating national power in this way utilizes the gamut of sources, instruments, and forms of power and achieves a higher degree of effectiveness than acting along a single line of effort or along multiple lines of effort discordantly.

The Russian soldiers in Crimea were described as "polite people" due to the way in which they conducted themselves while interacting with the Crimean people; rather than using violence in its physical form, the polite people relied upon the use of non-physical violence to impose their will upon the individuals with whom they were required to interact [15]. Conducting the military aspect of the operation in this way exemplifies the psychological violence inherent in an act of condign power to force compels submission. Having previously been conditioned to recognize the potential violence inherent within any interaction with the Russian military, or those portraying themselves as such by mimicking preconceived behaviour by those who would conduct a military operation, the citizens with whom the polite people interacted took advantage of the courtesy of being afforded some semblance of a choice to submit and opted to do so. Conducting the military aspect of the operation in this way afforded Russia greater flexibility in its options. By withholding the use of force as violence, Russia retained violence as an option if it became necessary to force submission through physical application of condign power, but because doing so had a greater probability of compelling a significant military repercussion from third parties that course of action proved less desirable. The implication that violence might occur proved sufficient for Russia to achieve many of its military objectives in Crimea, thereby exemplifying the condign power in its potential form and demonstrating how force can be applied psychologically without the use of violence in its physical manifestation.

4.7 Deny Everything, Admit Nothing, and Make Counteraccusations

On March 18 in an address to the Duma, the Russian President reported that Crimea and Sevastopol had been admitted to the Russian Federation. While the audience of this speech was physically located in the Kremlin, its content was directed at an international audience, primarily the United States and the West, and intended to justify Russia's actions while discouraging interference in what Russia considered its own domestic affairs. This speech proves significant because it reveals the purpose, method, and the desired end state towards which Russia had worked for over a decade, and after reiterating its same messages essentially declares victory in its endeavour.

The speech begins by citing the common heritage binding Russia and Crimea by emphasizing culturally significant religious and military events over the previous millennium, citing again the baptism of the Rus in 983 and the presence of the Black Sea Fleet in Sevastopol as events that have forged an identity shared by the peoples of Crimea and Russia. From this premise, the conclusion comes that "In people's hearts and minds, Crimea has always been an inseparable part of Russia. This firm conviction is based on truth and justice and was passed from generation to generation, over time, under any circumstances, despite all the dramatic changes our country went through during the entire twentieth century" [30]. This argument presents the status quo ante as the unnatural state of affairs, which if true would not only legitimize the Russian annexation of Crimea but also afford Russia the glory of being the liberator by acting in support of the Crimean right to self-determination and Russia's responsibility to protect those it considers diaspora [30]. Portraying itself in such light serves to distract from the fact that Russia justified its efforts to regain its lost province on the Black Sea and the unimpeded access to the global commons afforded by it under the guise of acting in accordance with the internationally recognized principles of self-determination and the responsibility to protect. The idea of common identity between the Russian and Crimean peoples becomes a conveniently useful means to an end, but a powerful one because it enables Russia to invoke self-determination and the responsibility to protect, two of the most critical principles shaping state actions in the twenty-first century international order. In support of this argument, the Russian President cites in the March 18 speech the Kosovo precedent set by the United States during the last decade of the twentieth century and the ruling by the United Nations International Court in 2010 [30]. Damning passages appear in the official position of the United States, which states that "Declarations of independence may, and often do, violate domestic legislation. However, this does not make them violations of international law," and the Court's response that "No general prohibition may be inferred from the practice of the Security Council with regard to declarations of independence," and "General international law contains no prohibition on declarations of independence" [35, p. 6, 36, p. 51]. The March 18 speech argues that Russia's actions comply with these rulings, which if true could be an attempt to compel the United States to choose between standing by the Court's opinion and not intervening in Russia's Crimean affair or violating its own arguments and undermining what little authority the Court possesses by staging some sort of intervention on behalf of Ukraine. Events have demonstrated that the United States opted for the former.

4.8 The United States Reacts

In response to the Russian annexation of Crimea, the United States exercised its informational and economic national power by publicly condemning Russia's actions and implementing sanctions, as well as providing Ukraine with armaments and combined military exercises intended to bolster the wavering Ukrainian military. The reaction of the United States provides examples of its national power in action and indicates how the United States reconciled its national interest with the international security dilemma, in terms of both acute consequences and long-term impacts, sparked by the Russian annexation of Crimea.

The United States recognized indications and warnings of Russia's intentions toward Ukraine as early as 2008, when analysis suggested that the increased momentum for Ukraine's membership in NATO could catalyse Russian activities in the region intended to prevent Ukraine from beginning the membership action plan: "Russia could try to stir up regional conflicts between eastern and western Ukraine...encourage pro-Russian groups to intensify anti-NATO campaigns, and stir up conflict by pushing for use of Russian as an official language in eastern and southern Ukraine. Russia could even encourage those favouring more autonomy for these regions or even their separation from Ukraine, particularly in the case of the Crimean Peninsula [7, p. 25]. As reviewed in previous sections of this chapter, Russia completed each of these activities and did so in an integrated, mutually reinforcing manner. Despite Gallis and the United States Congressional Research Service having predicted precisely how Russia would operate, the United States allowed conditions to develop that would prove insurmountable for Ukraine, its prospects for NATO membership, and the expansion of NATO and its Article 5 clause to Russia's southern border on the Black Sea [7, p. 25]. The loss of the initiative to Russia in Crimea forced the United States to exercise its power reactively from a position of weakness relative not only to the physical location of Russian military forces, but also and more importantly the perception of propriety that Russian information operations had given to its entire endeavour. In this way, loss of the initiative corresponded to losing the willingness to expend the requisite amount and type of power to revert conditions to the status quo ante. With its options limited, the United States reacted to the Russian incursion and annexation by using its national power to communicate its own narrative intended to counteract and discredit Russian information operations while simultaneously reducing Russia's capacity to generate and project power in a similar way again by targeting key organizations and individuals with economic sanctions.

One possible explanation for the limited response from the United States in this instance relative to previous episodes of invasions and incursions like the Iraqi invasion of Kuwait in 1990 arises from a lesser degree of association between the United States national interest and that of Ukraine and specifically Crimea [26]. The United States depends upon neither Ukraine nor Crimea as a significant component of its national interest because the national identity of the United States at the time of the Russian annexation does not mandate action beyond condemnation and sanctions.

In 1991 during the prelude and conduct of the Gulf War, the United States conceived of its identity in more militaristic and global terms as a consequence of having endured the stress of the Cold War and its use of the military as the primary means for deterrence. When an opportunity for limited action with international support for reinforcing norms presented itself with the Iraqi occupation of Kuwait, an alignment of national identity and interest facilitated the exercise of military power in pure conventional combat—notwithstanding the historic Iraqi claim of cultural sovereignty over Kuwait reflective of the Russian claim over Crimea: "In 1961, when the United Kingdom ended its protectorate over Kuwait, then Iraqi Prime Minister General 'Abd Al-Karim Qasim' asserted that Kuwait was an 'integral part of Iraq' because it had been part of the former Ottoman province of Al-Basrah [26]. The circumstances surrounding the Gulf War—that is, the national identity and interest of both the United States and Iraq as the primary belligerents—reflects the conditions that characterized the situation between Russia and Ukraine regarding Crimea and the United States, which by 2014 had redefined its identity and interest in the 2010 National Security Strategy. Whereas in 1991 the United States remained ascendant and perhaps achieved its zenith, the United States of 2014 had endured two protracted conflicts in Iraq and Afghanistan and the consequences of the global recession that began in 2008. The 2010 National Security Strategy redefined the national identity of the United States by focusing on domestic matters, or rather cultivating the sources of power by generating property and organizing it as a way of restoring the visages of personality reflected by United States foreign policy more akin to its period of ascendency that transpired towards the end of the twentieth century: "The welfare of the American people will determine America's strength in the world, particularly at a time when our own economy is inextricably linked to the global economy. Our prosperity serves as a wellspring for our power" [21, p. 9]. The national identity and interest of the United States as outlined in the 2010 National Security Strategy sought the preservation of international norms as a way of achieving "the political and economic benefits that come with greater integration with the international community" with the consequence for refusal to cooperate meted out as isolation from the boons associated with the globalized world and its economic potential [21, p. 11]. International engagement would rely upon integrating all aspects of national power:

> Diplomacy and development capabilities must help prevent conflict, spur economic growth, strengthen weak and failing states, lift people out of poverty, combat climate change and epidemic disease, and strengthen institutions of democratic governance. Our military will continue strengthening its capacity to partner with foreign counterparts, train and assist security forces, and pursue military-to-military ties with a broad range of governments. We will continue to foster economic and financial transactions to advance our shared prosperity. And our intelligence and law enforcement agencies must cooperate effectively with foreign governments to anticipate events, respond to crises, and provide safety and security [21, p. 11].

By emphasizing preparedness over response, the United States sought to prevent conflict from occurring as a way of avoiding further entanglements like opening additional theatres of military operations, and to mitigate the consequences of conflict where it does occur. The desired goal of this strategy—or, what the United

States wants to achieve and in so doing satisfy its self-understanding constituting national identity—would manifest as "a just and sustainable international order—just, because it advances mutual interests, protects the rights of all, and holds accountable those who refuse to meet their responsibilities; sustainable because it is based on broadly shared norms and fosters collective action to address common challenges" [21, p. 12]. A critical vulnerability appears between pursuing a strategy dependent upon norms but lacking the commitment to enforce norms, particularly those that have eroded due to new precedents: "International institutions must be more effective and representative of the diffusion of influence in the twenty-first century. Nations must have incentives to behave responsibly or be isolated when they do not" [21, p. 12]. The limited exercise of power by one agent sets conditions that favour another who places no limits on the use of instruments and forms of power to achieve its interest and by doing so satisfy its identity.

The United States attempts to enforce norms using economic power against the exercise of military and informational power that Russia used against Ukraine proved futile due to the mismatch between the extensiveness, intensiveness, and comprehensiveness associated with each national power domain. Military and information campaigns wielding the instrument of condign power using physical and psychological violence achieves a higher degree of extensiveness, intensiveness, and comprehensiveness than does the compensatory power of economic sanctions, even when combined with the conditioned power of persuasion afforded by public condemnation. The cost of annexing Crimea proved justifiable for Russia and its estimation that the international response would not produce a reciprocal response to its use of military force.

4.9 Too Little, Too Late

On February 28, 2014—the day after Russia military forces seized the Crimean Parliamentary building—the United States issued its return volley with the President issuing remarks directly contradicting the Russian messaging and did so using the same principles of self-determination and the responsibility to protect that Russia had used to justify its intervention. In his remarks, the United States President invoked the principle of self-determination twice, opening the statement by saying that "the Ukrainian people deserve the opportunity to determine their own future" and then concluding that "the Ukrainian people have also reminded us that human beings have a universal right to determine their own future" [22]. This line of argumentation follows the same logic that the Russian President used, with the critical difference being for the purpose of the argument; both Presidents affirm the self-determination principle, but the United States uses it to uphold Ukraine's sovereignty while Russia uses it to validate the referendum that would occur in Crimea one week later. While the United States invokes self-determination as a third party to the conflict between Ukraine and Russia on behalf of the former, the latter uses it to justify its own actions and satisfy its own interest. The critical difference is therefore at which level of organization the principle of self-determination emerges. The United States argument is that emergence of self-determination occurs at the state-level, and can only belong to the state, whereas the Russian perspective affords the right of self-determination to stateless peoples and entities devoid of sovereignty. Examples exist where both the United States and Russia have advocated for the opposite interpretation of self-determination, most notably in Kosovo.

The United States President also acknowledges the ties that bind Ukraine and Russia, even highlighting the Russian military presence in Crimea. This recognition echoes the Russian messaging that characterizes the relationship between Ukraine and Russia as special to give the impression that Russia may exercise certain privileges, such as the responsibility to protect ethnic Russians in Ukraine, despite Ukrainian sovereignty or any stringent interpretation thereof. While Russia argues that the commonalities between the Ukrainian and Russian peoples justify its actions, the United States decries such logical fallacy and holds that even though "Russia has a historic relationship with Ukraine, including cultural and economic ties, and a military facility in Crimea, any violation of Ukraine's sovereignty and territorial integrity would be deeply destabilizing, which is not in the interest of Ukraine, Russia, or Europe," and projects further the cascading impacts of the Russian precedent being set in Crimea as "profound interference in matters that must be determined by the Ukrainian people. It would be a clear violation of Russia's commitment to respect the independence and sovereignty and borders of Ukraine, and of international laws" [22]. The consequences of the Russian incursion into Ukraine and the annexation of Crimea directly violate established international norms, even if such a thing as international law could not be said to exist on account of the lack of enforcement mechanism available; the indirect consequences would be reflected in the future actions of Russia, or even the United States or another state, attempting to replicate these actions and arguments in their own interest.

In the following month, the United States would follow its initial rebuke with three executive orders issuing in tandem economic sanctions targeting the economic capabilities and capacity of the Russian Federation and accompanying explanatory statements intended to counteract Russian information operations, with each building upon the previous set. On March 6, Executive Order 13660 blocked the "transfer, purchase, export, withdrawal, or dealing in" property held by both Russian entities and individuals responsible, complicit, or otherwise involved in any activities "that undermine democratic processes or institutions in Ukraine; threaten the peace, security, stability, sovereignty, or territorial integrity of Ukraine; or misappropriate any state assets of Ukraine or of an economically significant entity in Ukraine," to include the assertion of "governmental authority over any part or region of Ukraine without the authorization of the Government of Ukraine" or any support to activities in pursuit of that goal [38]. This executive order also denied entry into the United States by individuals and donations to entities meeting the criteria for sanctions. The President of the United States described the executive order as authorization for "sanctions on individuals and entities responsible for violating the sovereignty and territorial integrity of Ukraine, or for stealing the assets of the Ukrainian people" and characterized this exercise of power as a first step, with the determination to apply additional pressure dependent upon the Russian counteraction [23]. The President stated that the United States enacted the sanctions to protect "the principle of state sovereignty" and declared that "the proposed referendum on the future of Crimea would violate the Ukrainian constitution and violate international law. Any discussion about the future of Ukraine must include the legitimate government of Ukraine" [23]. The remarks conclude by offering opening terms for negotiations between the immediate parties to the conflict as well as the international community, presumably led by the United States:

> There is also a way to resolve this crisis that respects the interests of the Russian Federation, as well as the Ukrainian people. Let international monitors into all of Ukraine, including Crimea, to ensure the rights of all Ukrainians are being respected, including ethnic Russians. Begin consultations between the government of Russia and Ukraine, with the participation of the international community. Russia would maintain its basing rights in Crimea, provided that it abides by its agreements and respects Ukraine's sovereignty and territorial integrity. And the world should support the people of Ukraine as they move to elections in May [23].

The offer presented in 2014 did not survive the presidential transition in the United States that occurred in 2017, indicating a major shift in United States foreign policy both philosophically and practically that receives further attention in chapter five. Crimea remains under Russian control with no indications of international motivation to undertake action seeking to reverse the annexation.

4.10 Never Bring Economic Sanctions and Harsh Rhetoric to a Gun Fight

The United States enters the fray at an insurmountable strategic disadvantage due to the way in which Russia has conducted its operation, which has spanned decades and integrated exercises of national power so that each mutually reinforces the others in both time and space. Russia precluded the use of military force, in this sense understood as violence and the tangible-physical application of condign power, against its paramilitary personnel active in Ukraine and especially Crimea by acting under the auspices of legitimate authorities and principles recognized across the international community, specifically self-determination and the responsibility to protect. The United States is left with little recourse, as it cannot justify the accompanying risk of any use of military force against Russian or Crimean units. Without the ability to remove the Russian occupation force in a physical sense, the United States essentially concedes the ground and resorts to exercises of power through the economic and information domains—though neither offer any real possibility of compelling a return to the status quo ante. This situation indicates the concession of Crimea to Russia by the United States and furthermore introduces the possibility that exercises of national power undertaken without a correlative condign element that both can introduce violence and has convinced the adversary of that very real possibility stand little chance of success. The determinative factor in cases such as this is cost. The United States chose to exercise economic and informational power because it was unwilling to repurchase Crimea for Ukraine in American—or Ukrainian—blood. Instead, the United States opted for a long-term approach measured in generations that relies upon the application of compensatory and conditioned power.

This model outlines the United States' attempt to exercise power in response to the Russian annexation of Crimea. The United States imposed economic sanctions against Russia and utilized diplomacy to condemn the action. The use of economic sanctions finds its source in property and works through the condign and compensatory instruments to apply force that seeks to compel a behavioural modification within Russia as the intended power subject; in this case, sanctions act primarily through the compensatory instrument with secondary influence from the psychological impact of the use of force via condign power. Public condemnation relies upon the state as organization to exercise persuasive conditioned power intended to mitigate future instances of repeated Russian behaviour while creating the option for reversion to the status quo ante. As Russia remains in control of Crimea despite these efforts, the United States has not achieved the effectiveness criteria that would categorize its reaction to the Russian annexation of Crimea as an exercise of power.

Economic sanctions serve as negative application of compensatory power in that the sanctions seek to disrupt transactions and established access to markets emanating from Russia and the key individuals politically and financially supporting the incursion into Ukraine and annexation of Crimea [18]. Sanctions are intended to punish by reducing capital flows and the ability to utilize capital in any way contributing to Russia's operations against Ukraine. The existence of financial transaction networks facilitates positive compensatory power for those entities party to the transactions, whether directly or indirectly; it is the exchange of goods and services that produces economic growth to the degree that states may utilize the resulting products [8]. The deliberate severing of such relationships therefore constitutes the negative application of compensatory power as a denial of the materiel and capital derived from economic exchange.

The sanctions also produce two secondary effects that can be viewed as manifestations of conditioned and condign power. The economic generation capacity of a state, in this case Russia, determines the amount of capital flowing through its domestic engines. Entities and individuals alike depend upon capital and become acquainted with a general concept of how much money will be available to them. This produces a normalizing effect through conditioning. Disruption occurs when reality does not meet expectations; the status quo changes, in real and psychological terms. Just as purchasing power changes, so does the understanding and expectation of the entity and individual. In this way, each becomes conditioned to a new set of circumstances that influence expenditure decisions. Conditioning resulting from a negative application of compensatory power to a significant degree might also constitute an exercise of condign power in its non-violent, psychological form [4]. If a loss of prestige by the state, entities incorporated under its authority, or the citizenry results from the application of sanctions and its associated consequences, the sanctions have demoralized those impacted. The recognition of the sanctions by Russia, the power subject in this instance, as causing less capital or esteem, or fewer rubbles and the anxiety resulting from the transition to renormalized circumstances, is unnecessary to meet the deliberateness property of an act of power; the United States, as power holder, has undertaken its actions with the express purpose of achieving the effects here described.

The information operations undertaken by the United States in support of its economic sanctions represent an attempt to characterize changes to the status quo by subnational entities claiming and using privileges reserved for the state as unacceptable. The sanctions serve as the punitive measure, while the information surrounding the sanctions serves to communicate a message intended to condition those receiving it that actions like those of Russia and Crimea have no place within the present international order. The counterpoint indicated by the actions of Russia and Crimea is that the present international order cannot abide for much longer in the twenty-first century without some adjustment to account for an international community bound by interdependence across the national power domains; that is, an exercise of national power in a single domain by one state can produce cascading effects across the other domains. With non-state entities like Crimea usurping authorities previously reserved exclusively for the state, the state cannot be said to possess a monopoly on the legitimate use of violence—even if it still possesses the greatest capacity for exercising power relative to its competition.

4.11 Same Old Song and Dance

The United States followed its initial sanctions and condemnation with more of the same. Rather than add additional exercises of power, the United States only increased the scope of the economic sanctions implemented in Executive Order 13660 and continued condemning Russian actions with same-styled rhetoric in official statements. This strategy limited its own effectiveness by failing to integrate exercises of power from the military and diplomatic national power domains, or to introduce new measures within the economic and information domains. Increasing singular lines of effort, as opposed to reinforcing the economic and informational lines of effort established by Executive Order 13660 with additional ones across each domain, failed to produce any momentum towards fulfilling the intent of the United States as articulated in the President's remarks during March 2014. By repeating its exercise of power without introducing any additional sources, instruments, or forms, the United States accepts that it cannot achieve its desired effectiveness but satisfies its obligation to act, which makes the economic sanctions issued by the United States less an exercise of economic national power and more so an exercise within the information domain; the difference between the two arises from disparate sources and instruments as well as effects. Economic power operates primarily based on property as its source and through compensatory power to mete boons and punishments through the various forms, while informational power derives from personality and organization—understood as foils for individualists and collectivists—and operates through conditioned power as either persuasion or manipulation. Lacking the ability to impose economic sanctions that would compel Russia to return Crimea to

Ukraine, the United States chooses to use economic sanctions to deliver a message, albeit weak and ineffective, that allows it to satisfy an obligation to act. Condemning the violation of international norms through rhetoric alone means next to nothing so the accompanying sanctions give the image of further and more serious action.

On March 17, the President of the United States signed Executive Order 13661 that expanded sanctions to include seven key leaders of the Russian government, specifically in the arms sector, and allowed for the application of sanctions to any other individuals or entities identified by the United States government as having contributed to the annexation of Crimea [39]. In the accompanying official statement, the United States President declared the Crimean referendum illegal according to international law and that "it will not be recognized by the international community" [24]. The remarks conclude with an offer of reconciliation that from the perspective of the United States seems mutually-beneficial for both Ukraine and Russia: "There's still a path to resolve this situation diplomatically in a way that addresses the interest of both Russia and Ukraine. That includes Russia pulling its forces in Crimea back to their bases, supporting the deployment of additional international monitors in Ukraine, and engaging in dialogue with the Ukrainian government" [24]. The only collateral offered to induce Russian entry to diplomatic negotiations comes from the "possibility" of constitutional reform in Ukraine, which appears hypothetical at best considering the offer originates from a third-party to the conflict.

Three days later on March 20, Executive Order 13662 further expanded the sanctions to include individuals and entities indirectly contributing to Russia's actions in Ukraine and Crimea by targeting those involved in "financial services, energy, metals and mining, engineering, and defence and related materiel" that "have materially assisted, sponsored, or provided financial, material, or technological support for, or goods or services to or in support of" anyone sanctioned under the previous Executive Orders [40]. The supporting rhetoric echoes the sentiments previously expressed—the illegality of the Crimean referendum and subsequent Russian annexation, the prospect of further or additional incursions westward by Russia causing additional sanctions by at least the United States, and the enduring possibility of diplomatic reconciliation that suits the interests of all parties (Obama, March 20, 2014). The juxtaposed justifications from the Russian and United States perspectives of their respective actions reveals a reliance on the same common principle—persuasive conditioned power manifesting as an argument that the right of self-determination justifies the actions of each that operate in the same way despite presenting conflicting positions; The level—either national, subnational, or supranational—at which peoples might invoke this right divides the two perspectives and makes them fundamentally irreconcilable. Stalemate ensues. The juxtaposition between Russia and the United States exemplifies the unity of opposites between the theoretical and practical application of this power theory—all exercises of power utilize the same fundamental principles understood as the sources, instruments, and forms of power, while the unique identity and corresponding interest of the agents wielding power inherently create the conflict that appears in and shapes the actions between agents and consequently the environment in which all exist. A common process exists that all agents use to determine identity, interest, and exercise power, but the process produces different results for different agents that creates or sustains conflict.

4.12 The American Legislation Culmination

The United States reaction to the Russian incursion into Ukraine and annexation of Crimea climaxes with the passage of the Ukraine Freedom Support Act of 2014, which the President signed into law on December 18, 2014. The act formalizes the policy of the United States to support Ukraine and oppose Russia's actions against it that had first been implemented via the series of executive orders. The act establishes prohibitions on the guarantee, insurance, or extension of credit connected to the export of goods or services; on procurement of goods and services; on provision of defence articles and services; on property and banking transactions; on maintaining correspondent or payable-through accounts by designated foreign financial institutions; and on investment in equity or debt held by designated Russian individuals and entities [37]. Russian entities Rosoboronexport and Gazprom receive specific mention, with the former sanctioned for its role in supporting the Syrian government and the latter threatened with contingent sanctions should it withhold natural gas supplies from NATO members or Ukraine, Georgia, or Moldova [37]. In addition to the punitive measures towards Russia, the act also provides Ukraine with certain goods and services, including the appropriation of funds for military assistance in the form of anti-tank and anti-armour weapons, crew weapons and ammunition, counter-artillery radars, fire control range finder, optical and guidance and control equipment, troop-operated surveillance drones, and secure command-and-control equipment as well as funds dedicated to "increasing the quantity of Russian language broadcasting into former USSR states in order to counter Russian Federation propaganda" [37]. Despite the significant provisions for waging war, the act explicitly forbids its use as an authorization for the use of military force thereby limiting its scope to the economic and information domains.

Despite the extensive and detailed authorities established under the act, its passage does not constitute an exercise of national power. The act serves as an enabling activity that facilitates the subsequent application of national power, specifically via the economic domain through its corollary instrument, compensatory power. The application of sanctions, or the implementation of the act once it became law, is the way in which the United States exercised its national power. By targeting specific individuals and entities that supported Russia's ability to project power both directly and indirectly, the United States limited its actions in terms of extensiveness but achieved a relatively higher degree of comprehensiveness and an exponentially higher degree of intensiveness. Russia's own estimate of capital lost due to the sanctions imposed by the United States is $1.1 billion at the end of 2018 and though Russian gross domestic product has waned since 2014, it cannot be definitively stated that the sanctions imposed by the United States, or even all the sanctions imposed against

Russia during this time, have diminished the Russian economy and by extension its ability to generate and project power [2, 33, 41, p. 6]. Russia has retained control of Crimea since its occupation began in 2014 and given no indication of relinquishing its position, indicating Russian willingness to absorb the consequences of economic sanctions as the cost to control Crimea [34]. If economic sanctions cannot be said to produce any discernible effect upon the Russian economy, then these sanctions cannot constitute an exercise of power in at least the economic domain. The only real consequence of these sanctions would appear in the form of prestige lost, which remains subjective and independent from a state's ability to use its national power in pursuit of its interests. By sustaining its position in Crimea, Russia has demonstrated the form of power produced by resisting an attempt by an aspiring power holder to impose its will upon its presumed power subject.

4.13 The Power to Resist and Persist

The Russian counteraction to the United States reaction has epitomized the power exercised by resisting. By holding Crimea and beginning to build or rebuild the infrastructure required to project Russian military power, Russia has endured the pressure applied via economic sanctions and public condemnation by the United States and western intergovernmental organizations. The ability to resist applied power becomes an act of power itself; that is, the ability of Russia to withstand the power exercised against it becomes its own exercise of power. Resistance against the attempt by the United States to stymie Russian efforts indicate that Russia plays the role of the power holder. While Ukraine is certainly one power subject in this conflict, the United States became a power subject as well on account of its failed attempt to compel Russian submission to the interest of the United States. Russia achieved its goals by gaining complete control over Crimea and in particular the city of Sevastopol, retaining control of its new province despite the sanctions levied against it, and stymieing the membership action plan for Ukraine's prospective entrance to the North Atlantic Treaty Organization.

The Russian strategy since 2014 has been simple—to hold Crimea and Sevastopol by enduring the outcry and sanctions until those entities opposing Russian actions have become conditioned to accept its control of Crimea and Sevastopol. Having resisted the economic sanctions and public condemnations by holding its ground and allowing the conditioning effect of time to elapse, Russia achieved a new normal by allowing the initial shock caused by its annexation of Crimea and disruption of the status quo to render a new state of affairs. To guarantee its access to Crimea while creating a means of further disrupting Ukraine's economy, Russia constructed the Kerch Strait Bridge to connect the road and rail transportation infrastructure of southern Russia to Ukraine. The bridge facilitates the assimilation of Crimea into the Russian Federation in a real sense as the linkage between the two became manifest with its construction, while the flow of commodities, goods, services, ideas, and people further solidify the revitalized connection between Russia and Crimea; the bridge also serves as a barrier preventing vessels from porting in Ukrainian cities along the Sea of Azov.

4.14 Conclusion

The Russian annexation of Crimea served two purposes—to provide Russia with unimpeded access to the global commons via the Black Sea and to stop the eastward expansion of NATO, particularly the application of Article 5 of its charter. The event horizon the entire Russian operation, which spanned all domains of its national power and applied them proactively, is measured in decades if not centuries. Ukraine, while the immediate party to the conflict, had little chance of resisting the incursion and none whatsoever of compelling Russia to abandon its gains; despite recognizing in 2008 the near inevitability of this conflict based on superfluous declarations of intent, the United States failed to set conditions that might dissuade a Russian incursion before it occurred. The actual response of economic sanctions and harsh rhetoric unsurprisingly failed to dislodge Russian forces from the seized territory because the United States exercised power in only two national power domains, reactively, against targets prepared to withstand the consequences, and with no clear pathway for achieving its desired end state of reverting to the status quo ante.

In a world where both violence and non-violence beget violence, exercises of power that employ violence afford the agent with the most advantageous course of action in reducing opposition in the immediate, acute engagement as the dead have no recourse—but the living do, which requires agents that use violence to consider the longer-term effects of its use beyond the elimination of an opponent. For the state, the conditions under which the use of violence offers a clear advantage must account for the cost measured in capital, lives, prestige, and the impact of these losses on future power generation and projection capacity. Killing offers an effective solution for reconciling competing interests between agents, but its use bears significant reciprocal consequences for subsequent interactions with others that while not party to the previous engagement in which violence decided the outcome did observe the way in which the survivor behaved. This dynamic bears a conditioning effect on the remaining population; that is, by using violence to end one intersubjective social power relation, an actor has also produced a conditioning effect that can modify the calculus by which others conceive of their identity and interest relative to the actor. In this way, the use of force through applied physical violence bears consequences for the power holder, subject, and those observing that may subsequently engage with either holder or subject. The use of physical violence by one justifies its use by another and rarely—if ever—offers those who utilize it impunity from the inevitable secondary consequences, which in the globalized world become amplified to such a degree that the amplification itself becomes an exercise of power itself. Within an egalitarian international community, any precedent for the application of violence can be adopted by another under circumstances that appear to be or can be made to seem the same. Justification for the use of military force by the state therefore resides solely within the information domain. While reality remains unalterable, perception is inherently vulnerable to exploitation. Violence remains a means to an end, but the international community has conditioned itself to detest its use to such a degree that it has become an unlikely option for the state to use without first creating a jus ad bello case by utilizing the other domains of national power as either ruse, guise, or both.

Russia determined its interest best served by regaining control of Crimea, so it set about doing so in a way that afforded it the greatest opportunity for long-term success. By propagating a narrative for decades that appealed to the fundamental aspects of Crimean identity, justifying its actions by citing precedents set by the United States and the West, and reinforcing its diplomatic machinations with physical and psychological violence, Russia succeeded in manufacturing its case for the incursion, annexation, and assimilation of Crimea into its Federation. The Russian President acknowledged the course of action taken, its many facets, and one source of discord within the contemporary configuration of the international order:

> After the dissolution of bipolarity on the planet, we no longer have stability. Key international institutions are not getting any stronger; on the contrary, in many cases, they are sadly degrading. Our western partners, led by the United States of America, prefer not to be guided by international law in their practical policies, but by the rule of the gun. They have come to believe in their exclusivity and exceptionalism, that they can decide the destinies of the world, that only they can ever be right. They act as they please: here and there, they use force against sovereign states, building coalitions based on the principle "If you are not with us, you are against us." [30].

The concept of American Exceptionalism and its use by the United States as justification for its actions within the international order serves as a theme that runs through each case study as a fundamental driver of American identity, interest, and the exercise of power. It is important to note that while the United States serves as the subject of this comment and perhaps also its target, the audience encompasses all states with which Russia sees as playing a part in its national interest. Rhetoric such as this depends upon persuasive conditioned power to convince its audience to accept certain facts as a way of modifying beliefs that may translate into new behaviour that more closely reflects the preferences of the orator. The significance of conditioned power upon beliefs and subsequent actions of both individuals and collectives reveals itself in both the Russian invasion of Ukraine in 2022 and the January 6 insurrection at the United States Capitol.

References

1. Bentzen N (2016) Ukraine and the Minsk II agreement: On a frozen path to peace? European Parliamentary Research Service. PE 573.951 http://www.europarl.europa.eu/thinktank/en/document.html?reference=EPRS_BRI(2016)573951. Accessed 7 Jan 2022
2. Chatzky A (2019) Have sanctions on Russia changed Putin's calculus? Council on foreign relations, New York. https://www.cfr.org/in-brief/have-sanctions-russia-changed-putins-calculus. Accessed 10 March 2021

3. Christakis T (2014) Self-determination, territorial integrity and *Fait Accompli* in the case of Crimea (November 3, 2014). ZaöRV/Heidelberg JIL Vol. 75 (1), 2015 at 75–100. SSRN: https://ssrn.com/abstract=2531321
4. Dubowitz M (2014) Iran: the psychology of sanctions. The National Interest. May 22, 2014. https://nationalinterest.org/feature/iran-the-psychology-sanctions-10511. Accessed 24 Nov 2020
5. Felgenhaur T (1999) Ukraine, Russia, and the Black Sea Fleet Accords. Princeton University, Woodrow Wilson School of Public Affairs. https://apps.dtic.mil/dtic/tr/fulltext/u2/a360381.pdf. Accessed 7 Jan 2022
6. Franklin B (1765) Join, or Die. https://constitutioncenter.org/blog/the-story-behind-the-join-or-die-snake-cartoon. Accessed 8 Apr 2021
7. Gallis P (2008) CRS report for congress: enlargement issues at NATO's Bucharest Summit. USG, Washington, DC. https://www.hsdl.org/?view&did=486312. Accessed 7 Jan 2022
8. Haass R (1998) Economic sanctions: too much of a bad thing. Brookings Institute. June 1, 1998. https://www.brookings.edu/research/economic-sanctions-too-much-of-a-bad-thing/. Accessed 24 Nov 2020
9. Hansen H (2020) Fallacies. In: Zalta EN (ed) The Stanford Encyclopedia of philosophy (Summer 2020 Edition). https://plato.stanford.edu/archives/sum2020/entries/fallacies/. Accessed 25 Nov 2020
10. Harris C (2014) When Catherine the great invaded the Crimea and put the rest of the world on edge. The Smithsonian Magazine; March 4, 2014. https://www.smithsonianmag.com/history/when-catherine-great-invaded-crimea-and-put-rest-world-edge-180949969/?no-ist. Accessed 7 Jan 2022
11. Ilarianov A (2014) Speech by Andrei Ilarionov at NATO Parliamentary assembly session in Vilnius. May 31, 2014. https://www.lrs.lt/intl/nato.show?theme=1060&lang=2. https://www.youtube.com/watch?v=J8ISQpbfoBI. Accessed 7 Jan 2022
12. Kaputsin A (2015) Crimea's self-determination in the light of contemporary international law. In: Symposium: "The incorporation of Crimea by the Russian Federation in the light of international law" Max Planck Institute for Comparative Public Law and International Law, Heidelberg, September 2–3, 2014. http://www.zaoerv.de/75_2015/75_2015_1_a_101_118.pdf. Accessed 7 Jan 2022
13. Kofman M, Migacheva K, Nichiporuk B, Radin A, Tkacheva O, Oberholtzer J (2017) Lessons from Russia's operations in Crimea and Eastern Ukraine. RAND Corporation, Santa Monica, CA. https://www.rand.org/pubs/research_reports/RR1498.html. Accessed 8 Apr 2021
14. Liang Q, Xiangsui W (1999) Unrestricted warfare. PLA Literature and Arts Publishing House, Beijing, February 1999. https://www.c4i.org/unrestricted.pdf. Accessed 1 Apr 2021
15. Lidz G (2015) Polite people of Russia not who you might expect. Newsweek, April 11, 2015. https://www.newsweek.com/polite-people-russia-321759. Accessed 7 Jan 2022
16. Marson J (2009) Putin to the West: Hands off Ukraine. Time; May 25, 2009. http://content.time.com/time/world/article/0,8599,1900838,00.html. Accessed 7 Jan 2022
17. Massie R (1980) Peter the great: his life and world. Random House, New York
18. Masters, J (1999) What are economic sanctions? Council on foreign relations. April 12, 1999. https://www.cfr.org/backgrounder/what-are-economic-sanctions. Accessed 24 Nov 2020
19. Michel C (2015) The Crime of the century. The New Republic; March 4, 2015. https://newrepublic.com/article/121222/one-year-after-russias-annexation-world-has-forgotten-crimea. Accessed 7 Jan 2022
20. NATO (2008) Bucharest Summit Declaration. https://www.nato.int/cps/us/natohq/official_texts_8443.htm. Accessed 7 Jan 2022
21. Obama B (2010) National security strategy. May 1, 2010. The White House, Washington, DC. https://obamawhitehouse.archives.gov/sites/default/files/rss_viewer/national_security_strategy.pdf. Accessed 5 Mar 2021
22. Obama B (2014a) Statement by the President on Ukraine on February 28. The New York Times, New York City. https://www.nytimes.com/2014/03/01/us/politics/transcript-of-obamas-remarks-on-ukraine.html. Accessed 7 Jan 2022

23. Obama B (2014b) Statement by the President on Ukraine on March 6. The White House, Office of the Press Secretary, Washington, DC. https://obamawhitehouse.archives.gov/the-press-office/2014/03/06/statement-president-ukraine. Accessed 7 Jan 2022
24. Obama B (2014c) Statement by the President on Ukraine on March 17. The White House, Office of the Press Secretary, Washington, DC. https://obamawhitehouse.archives.gov/the-press-office/2014/03/17/statement-president-ukraine. Accessed 7 Jan 2022
25. Obama B (2014d) Statement by the President on Ukraine on March 20. The White House, Office of the Press Secretary, Washington, DC. https://obamawhitehouse.archives.gov/the-press-office/2014/03/20/statement-president-ukraine. Accessed 14 Feb 2025
26. Office of the Historian, Foreign Service Institute, United States Department of State (2021) Milestones 1989–1992: The Gulf War 1991. https://history.state.gov/milestones/1989-1992/gulf-war#:~:text=Iraq%20accused%20the%20United%20States,United%20Arab%20Emirates%20and%20Kuwait. Accessed 5 Mar 2021
27. Pifer S (2020) Crimea: six years after illegal annexation. Brookings Institute. https://www.brookings.edu/blog/order-from-chaos/2020/03/17/crimea-six-years-after-illegal-annexation/. Accessed 22 June 2021
28. Pravda (1954) Presidium of the Supreme Soviet of the USSR, Decree of the Presidium of the USSR Supreme Soviet transferring Crimea Province from the Russian Republic to the Ukraine Republic. February 27, 1954. Original Source: Pravda, 27 February 1954, p 1. Available: Current Digest of the Soviet Press (Columbus, Ohio: AAASS, 1954), Vol 6, p 23. (Accessed 19 Mar 2019 via Seventeen Moments in Soviet History Project, Macalester College, Michigan State University. http://soviethistory.msu.edu/1954-2/the-gift-of-crimea/the-gift-of-crimea-texts/transfer-of-crimea/. Accessed 7 Jan 2022
29. Putin V (2013) Orthodox-Slavic values: the foundation of Ukraine's civilisational choice. Public Remarks in Kiev. http://en.kremlin.ru/events/president/news/copy/18961. Accessed 7 Jan 2022
30. Putin V (2014) Address by President of the Russian Federation on March 18, 2014. The Kremlin, Moscow. http://en.kremlin.ru/events/president/news/20603. Accessed 7 Jan 2022
31. Putin V (2014) Vladimir Putin answered journalists' questions on Ukraine on March 4, 2014. The Kremlin, Moscow. http://en.kremlin.ru/events/president/news/20366. Accessed 7 Jan 2022
32. Russian Federation (2014) Laws on admitting Crimea and Sevastopol to the Russian Federation. The Kremlin, Moscow. http://en.kremlin.ru/acts/news/copy/20625. Accessed 7 Jan 2022
33. Russian Federation Ministry of Economic Development (2019) The amount of trade restrictions lifted by the Ministry of Economic Development of Russia reached $330 million. http://economy.gov.ru/minec/press/news/2019021802. Accessed 7 Jan 2022
34. UAWire (2019) Kremlin: Russia lost over 6 billion dollars due to sanctions. February 19, 2019. https://www.uawire.org/kremlin-russia-lost-over-6-billion-dollars-due-to-sanctions. Accessed 7 Jan 2022
35. United Nations International Court of Justice (2010) Reports of judgments, advisory opinions, and orders accordance with international law of the unilateral declaration of independence in respect of Kosovo Advisory Opinion. July 22, 2010. https://www.icj-cij.org/public/files/case-related/141/141-20100722-ADV-01-00-EN.pdf. Accessed 8 Apr 2021
36. United States Government, Department of State (2009) Written statement of the United States of America: accordance with International Law of the Unilateral Declaration of Independence by the Provisional Institutions of Self-Government of Kosovo (Request for an advisory opinion). https://www.icj-cij.org/public/files/case-related/141/15640.pdf. Accessed 8 Apr 2021
37. United States Government. Congress (2014) Ukraine Freedom Support Act of 2014. USG Publishing Office, Washington, DC (PUBLIC LAW 113–272). https://congress.gov/113/plaws/publ272/PLAW-113publ272.pdf. Accessed 7 Jan 2022
38. United States Government. The President (2014) Executive order 13660 blocking property of certain persons contributing to the situation in Ukraine. USG Publishing Office, Washington, DC. Federal Register Vol 79, No 46. https://www.treasury.gov/resource-center/sanctions/Programs/Documents/ukraine_eo.pdf. Accessed 7 Jan 2022

39. United States Government. The President (2014) Executive order 13661 blocking property of additional persons contributing to the situation in Ukraine. USG Publishing Office, Washington, DC. Federal Register Vol 79, No 53. https://www.treasury.gov/resource-center/sanctions/Programs/Documents/ukraine_eo2.pdf. Accessed 7 Jan 2022
40. United States Government. The President (2014) Executive order 13662 blocking property of additional persons contributing to the situation in Ukraine. USG Publishing Office, Washington, DC. Federal Register Vol 79, No 56. https://www.treasury.gov/resource-center/sanctions/Programs/Documents/ukraine_eo3.pdf. Accessed 7 Jan 2022
41. Welt C, Archick K, Nelson R, Rennack D (2020) US sanctions on Russia. Congressional Research Service; Washington, DC. https://crsreports.congress.govR45415. Accessed 10 Mar 2021

Chapter 5
Diplomacy, Persuasion, and the Cost of the Paris Agreement on Climate Change

> Today the American people can be proud because this historic agreement is a tribute to American leadership. Over the past seven years, we've transformed the United States into the global leader in fighting climate change.
>
> Obama [22]

> The Paris Agreement handicaps the United States economy in order to win praise from the very foreign capitals and global activists that have long sought to gain wealth at our country's expense. They don't put America first. I do, and I always will...At what point does America get demeaned? At what point do they start laughing at us as a country? We want fair treatment for its citizens, and we want fair treatment for our taxpayers. We don't want other leaders and other countries laughing at us anymore. And they won't be. They won't be.
>
> Trump [29]

In December 2015 the parties to the United Nations Framework Convention on Climate Change (UNFCCC) agreed that climate change presents a global threat to every state in the community of nations and in response to "hold the increase in the global average temperature to well below 2 °C above pre-industrial levels and pursuing efforts to limit the temperature increase to 1.5 °C above pre-industrial levels" while also "fostering climate resilience" and "financing climate-resilient development" [32, p. 3]. Known as the Paris Agreement, this endeavour began in 1992 when 36 developed states—including several undergoing a transition to a market economy—created the UNFCCC and set its task as "to achieve stabilization of greenhouse gas concentrations in the atmosphere at a level that would prevent dangerous anthropogenic interference with the climate system" [30, pp. 9, 32–33]. Decades of diplomacy involving dozens of conferences and continuous negotiations produced the Paris Agreement, which became effective in October 2016 upon its ratification by 55 United Nations member states accounting for 55% of global greenhouse gas emissions [33]. For the United States, the period of negotiation from commencement in 1992 to ratification in 2016 spanned three presidential administrations of alternating political ideologies bookended by the advocates for the necessity of climate

K. L. Wedekind, *Constructing Power: A Critical and Constructivist Theory Using the American Case*, Advanced Sciences and Technologies for Security Applications,
https://doi.org/10.1007/978-3-031-84893-3_5

change adaptation as an element of its doctrine. Despite the expenditure of diplomatic capital and consequent progress towards the Paris Agreement, the United States would remain party to it for the least amount of time of any signatory. In November 2016, the United States voted into office an administration that would view the Agreement as a detriment to the interest of the United States, which would withdraw from the Agreement at the first available opportunity in 2020 [5].

This chapter examines the cost to the United States of championing the Paris Agreement for so long and then so quickly abandoning it by focusing on the expenditure of diplomatic power, the role of persuasion, and the effect of conditioned power during both the United States' advocacy for and retrenchment from the Agreement. The subject of climate change adaptation and the scientific evidence supporting its explanation will figure into this case study only in its relevance to the decisions and subsequent actions taken by the United States; that is, this chapter accepts the validity of the science that supports climate change and the consequent threat it poses. Following a brief review of persuasion and conditioned power, this case study will examine how the United States exercised both while leading up to the passage of the Agreement and the steps it has taken to withdraw from it.

The United States approach to the Paris Agreement facilitates an examination of how national identity and interest determine the way in which persuasive conditioned power operates through diplomacy. The relation between identity, interest, belief, and the exercise of power receives attention to illustrate the ways in which each impact the others. Personality and organization serve as the primary sources of power for the United States in its exercise of conditioned power using persuasion to advance international actions on climate change leading to the Paris Agreement in 2016, which illustrates how the process of organizing transpires and emphasizes the role that the four forms of authority other than legitimate authority play within an environment that does not recognize a legitimate authority. The United States' withdrawal from the Paris Agreement further demonstrates the links between identity, interest, belief, and power, and the consequences of retrenchment. Examining the role of the United States leading up to and since reaching the Paris Agreement of 2015 provides insight into the composition and exercise of persuasive conditioned power and its relationship with the identity and interest of the agent in question, the United States (Fig. 5.1).

This model utilizes process tracing to outline the exercise of persuasive conditioned power that the United States used in advocating for and opposing the Paris Agreement. The model offers a value neutral perspective on the exercise of power, which allows for the cultivation of power within the power theory presented in this work, which views the process as universal but the application of the products of the process as a unique reflection of the identity and interest of the agent wielding power. In this case study, the United States relies upon its personality and organization—understood as prestige and its diplomatic corps, respectively—to use conditioned power by persuading representatives from states party to UNFCCC through negotiations that rely upon the forms of authority to convince the others to modify their beliefs or at least behaviour by accepting the terms of the Agreement. These same principles—persuasive conditioned power buttressed by the four types of authority other

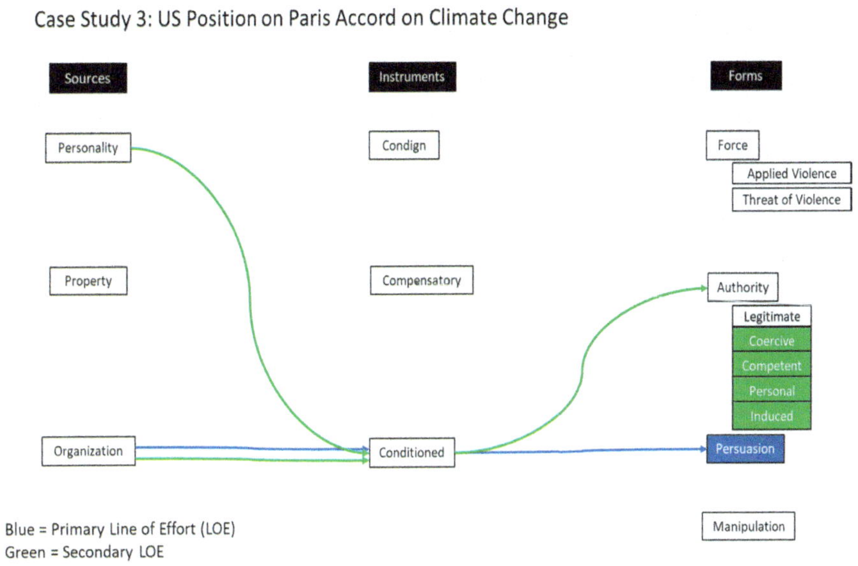

Fig. 5.1 US exercise of power in advocacy for the Paris Agreement (Author's own)

than legitimate—appear in the positions presented as justification for abandoning the Agreement following a change in administration within the United States.

5.1 Diplomacy as Persuasive Conditioned Power

Diplomacy comprises the formal negotiations between states and intergovernmental organizations and for its primary means relies upon the exercise of conditioned power, which consists of both persuasion and manipulation [14, p. 25]. Persuasion differs from manipulation, as the former transpires as a social relation wherein two or more entities exchange ideas and positions with the purpose of convincing the other to align its actions with the beliefs presented, while the latter occurs when one of the parties withholds its intent to convince and compel action that otherwise would not have taken place thereby denying the semblance of choice to the other party. Both forms of conditioned power may appear as mutually reinforcing lines of effort within an intersubjective social power relation, but due to their differences each constitutes its own power relation; that is, persuasion and manipulation do not prove mutually exclusive, but do operate independently from one another. Persuasion figures more prominently in the sort of formal diplomacy that produced the Paris Agreement than does manipulation due to the establishment of the UNFCCC and its investiture of legitimate authority by the UN member states to drive progress on international efforts to mitigate climate change. This act of organization serves as the source of power for the UNFCCC, while persuasion—one of the two forms of conditioned power—serves as the operative instrument within the UNFCCC and through which the parties to it could exercise diplomatic power against one another. While the goal of the collective might exist in terms of a singular purpose, the specific terms regarding how the individual members would contribute to achieving that goal remain contested until agreed upon. The negotiation process within the UNFCCC across its myriad conferences that produced the three primary objectives set forth by the Agreement transpired through the exercise of conditioned power in the form of persuasion [32, 35, p. 11].

Persuasion transpires as a duel of ideas between entities wherein each seeks to convince the other to modify its beliefs in some way. Parties engaged in this manifestation of the *zweikampf* will present premises and conclusions as an argument for consideration and either acceptance or rejection by the other [39, p. 3]. When one party accepts the validity of the position held by the other and alters the original belief, persuasion as an exercise of conditioned power has occurred. The entity whose ideas remain intact becomes the power holder, while the malleable entity takes the mantle of power subject. The holder obtains power when the subject accepts the arguments presented, which constitutes compliance with the will of the holder as opposed to the subject's own. Compliance in belief correlates to submission in action, whether that action manifests as thought, word, or deed [7, p. 142].

Within the context of the Paris Agreement, two forms of authority—competent and legitimate—play significant roles, but at different points and between different entities. The science that first identified the threat posed by climate change derives its validity from the scientists who produced it, specifically in the methods used to identify concrete evidence correlating human action to climate change and conclude that these changes do pose a global threat. The dependence between the validity of the methods used and the credibility of the scientists behind this conclusion inspires belief in others of its truthfulness. The measure of this belief in terms of extensiveness, intensiveness, and comprehensiveness directly correlates to the degree of competent authority possessed by the scientists and the actions taken by those who believe in their conclusions. The degree of belief held by individuals possessing legitimate authority within national governments, whether as the head of state or government, a cabinet member, senior policymaker, or diplomat, proves critical as it determines what actions the state might undertake to mitigate the threat posed by climate change. In this way, the competent authority wielded by scientists and their conclusions is limited by the corresponding trust and confidence vested into them by the legitimate authority holders; while competent authority derives its attributes from the belief placed in it by others, legitimate authority arises from an investiture by consensus [43, p. 51]. This distinction limits the extensiveness, intensiveness, and comprehensiveness of the competent authority due to the subjectivity of perception, as opposed to the codified, and therefore objective, limitations on legitimate authority. Competent authority possessed by the power holder varies based on the intensiveness of the belief held by the power subject. In the relation between science and government, the competent authority of the former depends upon the belief of the latter wherein persuasion, rather than validity, proves the determining factor.

The competent authority necessarily persuades its subject, as opposed to the legitimate authority that may exercise any form of power afforded to it by the source from which it comes, whether personality, property, or organization. Persuasion as a type of conditioned power finds its primary source in intellectual property, and to limited extent personality and organization corresponding to their respective ability to promote the intellectual property intended to exercise conditioned power through persuasion. This dynamic indicates a dependency and single point-of-failure for the competent authority that hinges on the belief of the power subject, which makes for a tenuous hold on power. When the power subject no longer recognizes the competent authority of the power holder, that entity no longer holds power. The case may also be that the power subject does accept the competent authority as such but acts otherwise, thereby denying the power holder a claim to competent authority, which renders it ineffective in persuading any potential power subjects to vest authority in it. A power subject might accept the competent authority of the power holder but fail to comply with its directives due to an inability to comprehend the requirements for compliance or a lack of skills; students failing an assignment due to a lack of requisite skills or sufficient time to complete an assignment exemplify cases in which competent authority remains intact, but the power subject does not comply.

The legitimate authority that empowers diplomats to negotiate on behalf of heads of state or government, and by extension the state itself, also relies upon the use of persuasion as conditioned power. The art of diplomacy at its essence is a negotiation of interests [20, p. 4]. Each diplomat seeks to secure for its state terms that facilitate the achievement of its national goals. The Westphalian principles of territorial integrity, non-interference, and egalitarianism endure as the framework that underpins the international order. While the first two characterize the concept of sovereignty, the idea that each state must treat the others as its equal, regardless of its national character or attributes, sustains conditioned power as the operative element in the relations between states; that is, in its purest form diplomacy as negotiation depends upon the egalitarian principle and its entitlement of equal standing among states within the international order. The absence of formal hierarchy makes the international order either anarchy or the purest manifestation of democracy yet; whichever the proper characterization, the egalitarian principle necessitates the use of persuasion and conditioned power as its primary means for official diplomacy, even while informal diplomacy, as constant companion, may utilize conditioned power in the form of manipulation, condign, or compensatory power.

5.2 All Roads Lead to Paris

The establishment of the UNFCCC in 1992 and its mandate to develop a way to mitigate the consequences of climate change began the process that culminated in the 2016 Paris Agreement. Progress occurred iteratively during formal diplomatic negotiation between states at conferences across the decades between commencement and conclusion, with key steps marked by the 1998 Kyoto Protocol, 2009

Copenhagen Accord, 2010 Cancun Agreement, and the 2012 Doha Amendment to the Kyoto Protocol—all of which facilitated adoption of the Paris Agreement, and the specific terms and conditions set forth in it [8, p. 8].

The charter that brought the UNFCCC in existence outlined the threat that climate change poses, the purpose of organizing to counter its effects, and the way in which the member states would set and pursue their respective goals to achieve the stated and agreed upon purpose. Recognizing the key premises necessitating collective action prove critical to the subsequent examination of how the United States would exercise its diplomatic power in support of and in opposition against the eventual Paris Agreement. The first admission in this series is the acknowledgement that the climate is in fact changing, and that these changes can have adverse effects, which human activities have exacerbated specifically through the production and release of greenhouse gases into the atmosphere thereby warming the surface of the earth and the atmosphere to damaging levels [30, p. 2]. The charter then distinguishes between the greenhouse gas emitted by developed and developing states, recognizing that while the former has contributed more to climate change thus far it is reasonable to expect the emissions of the latter to increase due to the necessary actions to further their economic development and social advancement [30, p. 3]. By making this distinction, the charter seeks to set reasonable expectations for states based on the unique circumstances that characterize their environment; doing so makes the charter itself transformational rather than transactional, as well as more realistic by accounting for feasible as opposed to idealistic possibilities. In an echo of the Westphalian principles, the charter recognizes that the unmitigated consequences of climate change would prove just as egalitarian by impacting every state, and reaffirms the sovereign rights of states to manage their domestic affairs, particularly the exploitation of natural resources as the means to pursue national interests pertaining to economic development and social advancement, but with the Hippocratic caveat that these activities do no harm to other states [30, p. 2]. This additional requirement sets national interest against the global cost, as each state must reconcile the benefit that it might receive against the cost in terms of environmental damage measured globally; that is, the collective of states represented by the UNFCCC proposes that the individual members ought to consider the impacts to the group when making their independent decisions, which remains their right as the sovereign entity. Underlying this approach is the implication that each state would determine their own standard related to the emission of greenhouse gases and those would be different due to the unique circumstances present within each state in terms of development, advancement, and interest. The intended distinction between developed and developing states appears as different expectations. Developed states, having relatively advanced economies and corresponding societies, would bear a greater share of responsibility in reducing global emissions because developing states have a requirement to use their resources "to achieve sustainable social and economic development and that, in order for developing countries to progress towards that goal, their energy consumption will need to grow" even if new technologies become available that might reduce the previous standards for emissions levels within which the now developed states operated [30, p. 9]. With the goals of the UNFCCC successfully achieved in the way set forth in the charter, the efforts of its members would have stabilized concentrations of greenhouse gases in the atmosphere "at a level that would prevent dangerous anthropogenic interference with the climate system" and done so quickly enough "to allow ecosystems to adapt naturally to climate change, to ensure that food production is not threatened, and to enable economic development to proceed in a sustainable manner" [30, p. 9]. The sense of urgency is emphasized no clearer than the caution expressed by stating that "where there are threats of serious or irreversible damage, lack of full scientific certainty should not be used as a reason for postponing such measures" [30, p. 9]. In the estimation of thc parties driving the establishment of this effort, the greater risk lies in delaying even in the face of uncertainty for its consequences seem graver than would an action taken with less than all of the information required to make the decision available at the time and incorporated into the decision-making process; that is, any sort of action offers a greater chance of mitigating the threat of climate change than nothing at all, and no excuses derived from what might seem like a lack of scientific information should delay the implementation of preventative measures.

A critical phrase appears in the charter that characterizes its purpose and the mission charged to its member states for implementation: "common, but differentiated responsibilities" (UNFCCC, p. 9). The purpose unites the collective, but each member must take account of its resources, capabilities, and interest and decide for itself the appropriate way to contribute to the overarching effort without neglecting its responsibility to either itself or the collective. Absent in this reconciliation of competing interests—what would become known as the "nationally determined contribution" in the 2016 Paris Agreement—is any official form of authority for validating the resulting goals for emissions [32, p. 4]. In place of such an authority, the UNFCCC and all its efforts would depend upon the member states to recognize that the threat posed by climate change presents a challenge that supersedes state interests by uniting them at the global level and compelling the submission of singular interests to those that benefit the collective; dependence on member states to determine their own contributions and implement domestic and foreign policies that will achieve these goals constitute a bottom-up approach [28, p. 3]. Compliance with the Paris Agreement depends upon the willingness of the members to comply with its stipulations and such willingness arises as the product of a conscious decision to participate or not. The decision-making process depends upon the propositions presented and considered, which derive from thc exercises of persuasive conditioned power that constitute diplomacy.

5.3 The Ambition of Kyoto

In 1997, the UNFCCC convened the third Conference of Parties (COP3) in Kyoto, which produced the first major endeavour undertaken in accordance with the UNFCCC. The Protocol comprised seven key stipulations, four of which introduced new concepts through which states could work to achieve the goals of the Convention:

developed, or industrialized, states agreed to reduce "emissions of six greenhouse gases by about five percent, on average, between 2008 and 2012, relative to 1990 levels [with] different national emission reduction targets" for each state known as "bubbles" (Article 4.1); developed states are allowed to trade the targeted national emissions rights, including bilaterally through a process known as "joint implementation" (Article 6.1); developed states can receive credit for emissions reductions that occur in developing states via the "Clean Development Mechanism" (Article 12.2); and the allowance of emissions credit banking, otherwise referred to as "international emissions trading" (Article 17) [16, pp. 1, 3]. The three additional provisions reaffirmed state sovereignty specific to domestic affairs undertaken in pursuit of the identified targets, parameters for accounting for "sinks or carbon sequestration focused on human-induced afforestation, reforestation, and retarded deforestation," and the requirement for ratification by 55 states before the Protocol would take effect [16, p. 3].

5.4 Bubbles, Joint Implementation, the Clean Development Mechanism, and Tradable Quotas, Oh My!

Each of the four instruments outlined within the Kyoto Protocol contrast its key provision reaffirming state sovereignty by threatening to erode the legitimate authority of the state over the conduct of its domestic affairs. The instruments set forth the ways by which the developed states identified in Annex B of the UNFCCC would work to achieve the goal of reducing "emissions of greenhouse gases by a certain percentage relative to its emissions in 1990. Each of these countries would have to achieve this reduction by 2008 and maintain the reduction through the compliance period 2008–2012" [15, p. 5]. The Protocol also prescribed the National Emission Reduction Targets, or bubbles, that each state would need to achieve individually for the collective to meet the goal. The goal for the United States set by the Protocol required 7% reduction of greenhouse gas emissions relative to 1990, which equated to 30% reduction of projected emissions for 2008–2012 [15, p. 5]. The problem with National Emission Reduction Targets proved two-fold. Firstly, the state would commit itself to goals set for it by the member states to UNFCCC and COP3, which would constitute an act of submission. The state would become power subject to the collective, in this case the UNFCCC and COP3, as power holder. Secondly, the developed states listed in Annex B of the Protocol operated under a market economy or were in the process of transitioning to a market economy in the case of the former Soviet states, to include the Russian Federation. While the state holds sovereignty over its domestic affairs from the perspective of the international community, market economy states limit their authority over the activities of private industry to facilitate growth and development dictated by supply and demand. Even if a state acquiesced to the goal set for it by the Protocol, the state might not possess the corresponding authority to curb emissions produced by certain sectors of private industry depending on the extent of domestic regulatory legislation.

With each market economy operating under unique circumstances and varying degrees of authority, the variable ability of each state to achieve the goals set for it necessitated the additional three market-based instruments as potential solutions. Article 6 of the Protocol attempts to alleviate the burden levied upon the state pertaining to private industry by allowing for "project-by-project exchanges among private entities in Annex B countries" amounting to a "cap-and-trade concept," which is defined as the establishment of "an allowable overall level of pollution and [its allocation] among firms in the form of permits, which can be freely exchanged among sources" [16, p. 5, 26, p. 20]. To enable cap-and-trade, national governments set a limit on greenhouse gas emissions by issuing emissions permits capped at specified levels to private entities, but also allow for the selling or trading—described as "transfer and acquisition" in the Protocol—of unused permits between private organizations in the developed states identified by Annex B [31, p. 6]. The cap-and-trade approach regulates the total emission output by the state, which is responsible for correlating the permits issued with its goal under the Protocol through annual reductions to the cap [17]. Under this arrangement known as joint implementation, private entities may trade permits domestically and internationally, which introduces some degree of flexibility in pursuing the goals of the Protocol as "the allocation of national targets affects the distribution of costs among nations" [15, p. 16].

The second mechanism that provides flexibility to developed states is the Clean Development Mechanism (CDM), which "gives emission reduction credit to firms in an industrialized nation for the reductions that occur as a result of its financing of an investment project in another country" [15, p. 24, 16, p. 5]. Enforcement proves difficult for the CDM as it seeks a relative reduction of emissions within the developing state, which may not have account of its national annual emissions level. Without such a baseline, it becomes difficult to say with certainty whether a project has reduced overall emissions or not.

Article 17 of the Protocol permits the trading of National Emissions Reduction Targets between developed states known as International Tradable Quotas [16, p. 5]. States that issue emission permits that go unspent can trade these to other developed states that may exceed their authorized emissions limit for a given year, which creates value for trading emissions credits and introduces an efficiency mechanism into the international emissions trading market. The Protocol explains that trading quotas in this way should augment domestic policies implemented by each Annex B state while providing flexibility by allowing for expansion or contraction of the emissions reduction goal set for it.

The three flexibility mechanisms identified by the Kyoto Protocol depended upon a flawed reconceptualization of state behaviour that doomed its ratification and subsequent implementation. The vision underlying the Protocol presumes to "translate non-cooperative, self-interested behaviour into collective cost-minimization when the participants in the trading system are profit-maximizing or cost-minimizing agents operating in a competitive market" [2]. The problem is that states "are not cost-minimizing agents, and even if they were, they tend to lack the information required to make such cost-effective trades" [16, p. 5]. Even if the state could collect and process the required data to make national strategy decisions about its annual emissions output, the only logical option available to the state would be to maximize its allowable emissions expenditure because doing so affords the greatest likelihood of economic growth—perhaps one of the few universal aspects of national interest shared by each state. The Protocol wrongly imagines that states would collaborate to counteract a common threat as opposed to continuing to seek the most advantageous conditions for itself; that is, the state and its interest remain not only sovereign but also supreme over all other considerations [16, p. 5]. The Protocol also depends upon developed states exercising legitimate authority to introduce new regulations or make existing ones more stringent. While such measures are within the scope of its authority to do so, implementation of additional regulation would alter the status quo, which could stymie economic growth until technology could adapt to the new requirements and mitigate any stagnation.

Expecting the state to slow its domestic economic growth on behalf of international goals is unlikely to align with its immediate economic goals and therefore in contrast with the interest of the state in the near term due to its redefinition that follows a change in political leadership; the shift in priorities within the United States between administrations, especially following the 2016 general election, demonstrates this point. The state, having organized and formed an identity reflective of its constituent parts, must seek economic growth as a way of sustaining itself financially by creating—or at least setting conditions that may create—property as a source of power that makes the exercise of power possible. It remains in the national interest of every state to increase its economic output in a way that sustains the state, which includes oversight mechanisms to keep growth at the most desirable rate for the given state. Expecting that the state would modify its economic growth for a potential long-term benefit is the critical flaw of the Protocol because it depends on a common understanding, shared belief, and willingness to act in the interest of all rather than oneself—all sustained across different Presidential administrations with contrasting identities and therefore competing interests. The election of a Presidential administration that opposes international collaboration in the United States in 2016 brought about a change in American national identity that led to the departure from the Paris Agreement.

5.5 An Unenthused Reception

By 2011, the Kyoto Protocol had been ratified by 192 United Nations member states—all except the United States, which had opposed the Protocol since the conclusion of COP3 as indicated by the passage of Senate Resolution 98 in July 1997 by a vote of 95-0 [1, p. 1]. The resolution cited two principal reasons for rejecting the terms of the Kyoto Protocol. The first justification argues for introducing "legally binding emission limitations targets" and associated timeline for developed states while placing no restrictions on developing states, for which the collective emissions forecasts at the time far exceeded those of developed states, would not sufficiently mitigate global greenhouse gas emissions [40, p. 2]. By allowing developing states to proceed unabated, the economic costs borne by developed states in pursuit of their emissions reduction targets would not lessen the global threat posed by climate change and therefore proved unjustifiable for the United States and counter to its interest. The second justification builds upon the first by stating that "the disparity of treatment between Annex I Parties and Developing Countries and the level of required emission reductions could result in serious harm to the United States economy, including significant job loss, trade disadvantages, increased energy and consumer costs, or any combination thereof;" the resolution concludes by declaring that the United States should not ratify any protocol that does not require all states to contribute to the effort or that would harm the United States economy [40, pp. 3–4]. Analysis produced by the administration and released to the United States House of Representatives Commerce Committee in 1997 identified the potential loss of 900,000 to 1.5 million jobs by 2005 if the United States were to undertake emissions reductions without corresponding action in developing countries [1, p. 4]. The refusal of the United States to ratify the Kyoto Protocol indicated to the international community that national interest superseded global interest and prevented any significant implementation of its parameters [10, p. 4].

By rejecting the Kyoto Protocol, the United States exercised its legitimate authority over the conduct of its domestic affairs. Complying with the terms set forth by the Protocol would require subordinating the interest of the United States not to the collective interest of the community of states but to the several interests of developing states, each of which would benefit from the implied permission to emit greenhouse gases with impunity. Exercising legitimate authority can come at a cost in terms of personal authority, understood in this case as prestige or standing within the community of states, if the United States were to lose its ability to influence the actions of other member states. Conversely, if this exercise of legitimate authority by United States inspired others to emulate its rejection of the Protocol then the United States has exercised coercive authority in addition to legitimate authority while expending personal authority. The critical difference between these three types of authority is that legitimate authority exists objectively as a structural property organizing the international order, whereas coercive, induced, personal, and competent authority are derived subjectively by the other party to a power relation and vested by the power subject into the power holder. The persistence of challenges to legitimate authority and the state wielding it—notably by violent non-state actors, transnational criminal organizations, among others—does not undermine legitimacy and may enhance it through othering; challengers to the status quo that derive their identities and define their corresponding interests relative to the state posture themselves as alternatives to the state, which implies and reaffirms its legitimacy. While the other four forms of authority prove relatively transient as perspectives and opinions ebb and flow, legitimate authority abides because the source of its power emanates from the act of organizing that provides the framework for the existence and functioning of the international order; that is, legitimate authority serves as the fundamental type of power necessary to distinguish between domestic and international affairs, which facilitates the existence of both. This distinction also serves to separate what exists from what appears to exist—the difference between reality and one's perception of reality. In the context of the Kyoto Protocol, this separation was exemplified by the delegate representing the United States signing the Kyoto Protocol in the United Nations headquarters as COP4 occurred in Buenos Aires in 1998, even though the United States Senate had passed Resolution 98 the previous year and the then President had stated his intent not to introduce the Protocol for ratification [16, pp. 2–3]. The signing of the Protocol in 1998 meant nothing because the pen of the delegate did not carry the legitimate authority of the United States to implement the Protocol, making the exercise a symbolic one intended to lessen temporarily the blow to the personal authority ascribed to the United States by the other delegates and by extension the states that each represented.

5.6 The Illegitimate Forms of Authority

Legitimate authority derives its power from the consensus of the individuals and entities subject to it through an act of collective submission to the power holder. Through this action, the subjects invest power in the holder knowingly and willingly, having decided that doing so serves the interest of the subjects. In power relations where legitimate authority does not exist, the other four forms of authority appear and serve as the means through which the parties to the power relation seek to dominate the others. The terms set forth by the UNFCCC charter created precisely such an environment—one wherein no legitimate authority exists, which affords the opportunity for the signatories of the charter to employ the other four types of authority either singularly or in combination with the others against each other and all United Nations member states [6, p. 2]. Coercive authority manifests as demonstrations undertaken and intended to deliver non-verbal messaging seeking to convince those observing that the presenting state possesses resolve beyond recourse to use its capabilities at whatever capacity might be required to compel compliance with its goals; coercive authority relies upon the exercise of condign power and the use of force in its non-physical or psychological form to achieve the submission of the intended power subjects. Induced authority transpires as an exercise of compensatory power when one state compensates another in exchange for its compliance. Induced authority represents the reciprocal of coerced authority; where the former bestows boons, the latter threatens punishment [43, p. 44]. Personal authority derives its power from prestige and as such depends upon perception; when one state seeks to gain favour based on its perception of another state by following its instructions or lead, then the power holding state wields personal authority [43, p. 60]. States, possessing qualities of both collectives comprising myriad entities and individuals as members of the international community, may exercise personal authority if and only if another state acts based on the perception of the existence of national character, which might exist as perceived in the minds of the decision-makers holding legitimate authority in rival states. Competent authority, having received its due previously in this chapter, cannot exist as a quality attributable to the state, as it derives its power from proficiency or knowledge that necessarily belongs to an individual as opposed to the perception of such, which is the sole distinguishing, and limiting, characteristic from personal authority [43, p. 63].

The illegitimate forms of authority—those other than legitimate authority—prove significant in the process of organizing individual agents into a collective. The international negotiations on climate change illustrate the organization formation process in that there does not exist any legitimate, consensus-based authority consolidated in one of the individual states. The negotiations seeking to establish an international climate change regime transpire using the four alternative authorities; persuasive conditioned power through the exercise of diplomatic national power between states provides the mechanism for organizing member states in a way that aligns their beliefs and consequent actions that manifest as part of their respective national identities and interests. The inability of discourse in the United States to overcome scepticism on the threat posed by climate change remains a significant detractor from international progress on climate change mitigation and its associated consequences. Within the United States, the populace remains divided regarding the issue of climate change, including belief in the science that identifies human activity as its source and the threat posed by climatological trends [13, p. 2]. Belief determines actions—including the way in which individuals vote in the United States. Disbelief in climate change domestically translates into reluctance to act internationally, which reflects the national identity of the United States. Sharp contrasts between presidential administrations and their respective policy priorities reduces the ability of the United States to wield the illegitimate forms of power effectively with respect to any issue within the typical timelines for international change, which for the issue of climate change mitigation has transpired over decades. Identity therefore determines interest and how the United States uses its power—either to advance or stymie—issues like climate change that require international cooperation.

5.7 From Champion to Dissident

In December 2009, Copenhagen hosted the fifteenth annual Conference of Parties (COP15) to the UNFCCC. Earlier that year, the United States inaugurated a new President who had identified climate change mitigation as a top foreign policy priority of his administration, as noted in a speech at the United Nations General Assembly in September that identified climate change as one of the greatest threats to global stability, which reclaimed a role for the United States as a global leader on climate change mitigation that had been scuttled since the refusal to ratify the Kyoto Protocol in 1998 [21, pp. 1, 4]. The Copenhagen Summit provided the first opportunity to initiate the corresponding action to accompany the revitalized policy goals of the United States but had begun to falter due to disagreement not on the threat posed by climate change but rather how the states—both collectively and individually—would mitigate the threat [19]. Even after direct intervention by heads of government from the United States, China, India, South Africa, Brazil, and the European Union, COP15 produced no tangible result, but it did succeed in three critical areas by establishing climate change policy as "central to the political thinking of every country on the planet," increasing public awareness of the issue due to messaging by "governments, NGOs and business and the media coverage of the issue and the summit have made addressing climate change widely understood and discussed from the pubs of rural England to the bars of Beijing," and legitimizing "green growth" as a viable economic development concept by refuting "the idea that addressing climate change is bad for business" [3]. While COP15 did not produce a binding agreement with specific emissions reductions targets, it demonstrated the willingness of states, and especially that of the United States, to negotiate such an agreement. Within the United States, recognition of the threat posed by climate change led to the development and introduction of a congressional resolution for a Green New Deal, which seeks to achieve net carbon neutrality within the United States by 2050 through a combination of technological and social change [38]. Disagreement within the United States Congress reflects the scepticism across the United States populace concerning the threat and consequences of climate change, with much criticism focused on the cost of implementation in the near term that neglects the potential long-term costs of climate change [12].

In the years following COP15 and preceding COP21 in Paris, the United States released three major policy documents, the Quadrennial Defense Reviews in 2010 and 2014 in addition to the 2013 President's Climate Action Plan that carried forth the momentum that it had regained in Copenhagen towards an international climate change agreement. Each of these documents present arguments intended to justify the necessity of designing and implementing new policies and programs that would mitigate the consequences of climate change. These justifications appeal both to a moral obligation to act and to the idea that doing so will prove in the interest of the United States by sustaining its military and economic advantages over its competitors. By the nature of these arguments, these documents target the perceptions and beliefs of multiple audiences, including the citizenry of the United States and its private industry as well as the international community, and seek to persuade any dissenters that action on the matter of climate change is not only justified, but in the interest of all—both individually and collectively. Presenting the mandate for climate change in an official United States strategic policy document therefore constitutes an exercise of persuasive conditioned power.

The 2010 Quadrennial Defense Review (QDR) describes the interest of the United States as "inextricably linked to the integrity and resilience of the international system. Chief among these interests are security, prosperity, broad respect for universal values, and an international order that promotes cooperative action" [36, p. iv]. The QDR also identifies climate change as having the potential to exacerbate conditions that cause conflict. While neither the single source nor direct catalyst of conflict, "climate change could have significant geopolitical impacts around the world, contributing to poverty, environmental degradation, and the further weakening of fragile governments. Climate change will contribute to food and water scarcity, will increase the spread of disease, and may spur or exacerbate mass migration," which "may act as an accelerant of instability or conflict" [36, p. 85]. The alignment of the interest of the United States as an individual state with collective interest of the global community—specifically related to climate change and its consequences for peace, stability, economic development, and social progress—would compel the United States to exercise its national power to fight climate change and its symptoms that might manifest as conflict driven by resource scarcity, pandemic, migration, poverty, environmental degradation, or failed and failing governance structures [36, pp. 86–87].

In 2013, the administration then in office issued the President's Climate Action Plan (PCAP), which justified the necessity of increasing the resources dedicated towards mitigating climate change and its effects in both moral and realist terms. The Plan argues that the United States possesses "a moral obligation to future generations to leave them a planet that is not polluted and damaged" and contends that such an obligation compels the United States to lead international efforts by both setting an example in its domestic policies and driving international negotiations within the UNFCCC towards the production of an agreement that incorporates all major states in an inclusive but flexible manner [11, p. 4, 16, 20]. To reinforce this appeal to morality, the PCAP presents its case on practical grounds by citing the consequences of climate change in terms of public health and economic costs:

> About one-third of all Americans experienced 10 days or more of 100-degree heat. The 12 hottest years on record have all come in the last 15 years. Asthma rates have doubled in the past 30 years and our children will suffer more asthma attacks as air pollution gets worse. And increasing floods, heat waves, and droughts have put farmers out of business, which is already raising food prices dramatically. These changes come with far-reaching consequences and real economic costs. Last year alone, there were 11 different weather and climate disaster events with estimated losses exceeding $1 billion each across the United States. Taken together, these 11 events resulted in over $110 billion in estimated damages, which would make it the second-costliest year on record [11, p. 5].

The combination of moral and realist justifications strengthens the case for action intended to mitigate the effects of climate change by increasing the number of ways in which persuasion as conditioned power can take effect on the targeted power subject. The multiplicity of arguments affords the power holder with more options and therefore a greater probability of engaging the power subject on an issue that figures into the identity and interest of the subject, which creates the opportunity for the holder to exercise persuasive conditioned power within the intersubjective social power relation. The use of persuasive conditioned power in such an instance indicates that it manifests as value neutral relative to the other forms and instruments of power. The decision afforded to the potential power subject accounts for the value neutral characteristic of persuasion, while the intentionality underlying the motives of the power holder makes it some measure less than truly neutral [43, p. 32].

The benefit of leading the international advocacy for action on the matter of climate change comes from the corresponding influence—understood in this context as conditioned power dependent upon a combination of coercive, personal, and induced authority—that the leader wields in determining the terms of action and the degree to which they bind the states subject to them. Conversely, rejecting the competent authority of climate scientists and their research coupled with the abandonment of an agreement on climate change would preclude any state so doing from influencing the community of states still dedicated to action; that is, influence wanes when an individual or entity, in this case the United States, removes itself from a group united by common purpose. Even if fundamental questions of validity persisted, remaining even if only to exert influence offers a more advantageous position than departing and attempting to exercise power over any outcomes from outside of the group. Stated as simply as possible, it proves more difficult to exercise power against a group than it does within that same group. An example of this principle in action arises from the comparison between the ability of a state to influence an alliance as an aspiring member and then as a full-fledged member; similarly, the ability of a state to influence the decisions made by the United Nations Security Council as a temporary member as opposed to its corresponding ability when not assigned to the Council. Given that intersubjective social power relations transpire between only two entities, these cases exemplify relations between an individual and a collective as opposed to relations between only individuals or only collectives. Examining the dynamics of relations comprising an individual and a collective would prove fruitful further research.

The 2014 Quadrennial Defense Review reaffirms the 2010 version and the PCAP by presenting the menace posed by climate change in terms of its global reach and emphasizing that its consequences could catalyse and exacerbate trends already stressing economic and social systems in both developed and developing states. Citing "water scarcity" and consequent "increases in food costs" as catalysts, the document offers that "the pressures caused by climate change will influence resource competition while placing additional burdens on economies, societies, and governance institutions around the world. These effects are threat multipliers that will aggravate stressors abroad such as poverty, environmental degradation, political instability, and social tensions—conditions that can enable terrorist activity and other forms of violence" (37, p. 8). The argument here presented in favour of the United States taking action to mitigate the effects of climate change aligns the interests of the United States and international community by recognizing that in a world characterized by connectedness, population growth, and development the effects of a global phenomenon will cascade across global society along geopolitical, economic, and social lines [18, p. 7]. As an argument, it seeks to convince its audience to adopt its perspective as its own and therefore constitutes an exercise of persuasive conditioned power, which proves significant because of the necessity for public support that translates into votes for candidates during election years that will sustain the endeavour to combat the effects of climate change—an inherent challenge associated with "environmental policy [that] is generally exposed to a severe and under-analysed generational conflict: it is not easy to forgo tangible present benefits to avoid distant and abstract detriment that will not meaningfully affect any of the current significant voting blocs" [4, p. 7]. This contest manifests in an action-reaction-counteraction cycle that transpires between 2015 and 2017 wherein COP21 produces a global agreement, the United States citizenry elects an administration that campaigned in opposition to such an agreement, and the United States announces its intent to withdraw from the Paris Agreement. A change in personality, one of the sources of power, appears as a shift in national identity that ripples across national interest and the endeavours for which the United States as agent is willing to exercise power, or not.

5.8 The Paris Agreement

The Paris Agreement offers an international approach to climate change adaptation characterized by differentiated balance. The Agreement relies upon both voluntary and mandatory aspects in the form of nationally determined contributions and the corresponding reporting mechanism that states would use to facilitate accountability across the signatories to COP21. Requiring each state to determine its own contributions avoids contrasting with the Westphalian principles constituting state sovereignty while still setting a specific, albeit non-binding, target for emissions reductions; in this way, the nationally determined contributions prove voluntary. The mechanism through which states would report progress towards the self-set goals would apply equally to all states. By combining the decentralized concept of nationally determined contributions with the centralizing reporting mechanism, the Agreement achieved some degree of balance. The differentiation manifests in the various targets set by each state for itself. This aspect reduces the friction cultivated by the idea of separate requirements for developed and developing states outlined in the Kyoto Protocol, which led to its demise. The chief negotiator from the United States delegation to COP21 described the Agreement as "a legal hybrid, blending binding elements of accountability with non-binding emissions targets. In critical ways, it bets on the force of rising norms and expectations rather than law to achieve its aims" [27, p. 3]. The success of the Paris Agreement therefore depends upon neither legitimate authority as in the law nor condign power as in the enforcement mechanisms possible under law, but rather in the several forms of illegitimate authority and persuasive conditioned power—all of which depend upon subjective beliefs that inform both the identity and interest of the involved parties. Comparative authority manifests in consideration of the form of power that occurs when an entity acts due to the reconciliation of reality with its ideal form or desires, which causes its subsequent action to rectify the deficiency sustaining the distinction; in such a form of power, the intangible aspirations of the future levied upon the undesirable present state necessarily transpire within a single entity. If such a phenomenon were to occur within a collective, then it could be said to possess the characteristics of an individual unit—most notably of which is a degree of bimodal symmetry significant enough to transcend the abyss between an individual and collective organization possessing the qualities of an individual.

5.9 The Shortcomings of the Agreement and Diplomatic Power

Critics have cited the Paris Agreement as fundamentally flawed due to the misalignment of the proposed solution with the problem, arguing that the magnitude of the forecasted effects of climate change have surpassed the rectifying capacity of any sort of diplomatic agreement: "deep cuts in emissions of carbon dioxide (CO_2) and other pollutants, as required ultimately to stop planetary warming, requires transformation of energy systems that central architects and standard diplomatic procedures cannot orchestrate" [41, p. 4]. One supporting criticism cites "the shifting interests of organized political constituencies, including voters" and the way in which their influence "alters how government behaves and how those shifts in behaviour alter the content of both policy and diplomacy" [41, p. 4]. The capriciousness that prevents diplomacy from succeeding on the matter of climate change adaption derives from the misalignment of national interest and climate change goals: "For nearly all countries and firms, climate change goals are not a central driver of change in energy markets and geopolitics. Instead, the reverse is true—protecting the climate is one of many policy goals often buffeted by larger trends in political interests and technology" [41, p. 2]. The primary drivers for action in the pursuit of interest come from determinations on what goals once achieved provide the state with the most advantageous conditions for its own success and aligns its interest with that of the collective only when the interest of the collective happens to align with the state's conception of its own interest. Alignment occurs by happenstance rather than through recognition that the interest of a single state might be best served through reconceiving its interest in a way that facilitates international collaboration—especially on issues like climate change that transcend the borders of legitimate authority owned by the state. Diplomacy serves as a way to overcome the political and economic inertia that prevents states from synchronizing their interests, which requires "Top-level political leadership [to] make a difference in shaping societal and global expectations—as [the then United States] President did with sustained high-level political engagement on the climate question, or as the French government has done in the run-up to the Paris Agreement and its aftermath" [41, p. 2]. These efforts begin as a priority for a head of state or government and reverberate in public policy and communications, and only begin to take effect when implemented as ideological goals for the state, which then drives its exercises of national power in the diplomatic domain. The advent of technologies that bolster arguments in favour of climate change action would increase the effectiveness of persuasive conditioned power on the matter. Diplomacy without the "practical deployment of new technologies [to] drive changes in emissions, political preferences, and how national governments assess national interests" has and will continue to fail in convincing states to undertake seemingly drastic measures to combat climate change that appear counter to current conceptions of national interest [41, p. 2]. The significance of how national interest impacts governmental action—whether unilaterally or multilaterally—on climate change adaptation cannot be understated. For every state, national interest changes to reflect the political leadership in office at a given point in time, and even within the duration of a single administration the priorities associated with the identified national interest change. The lack of consistent priorities dilutes the corresponding allocation of resources, which ultimately stymies progress on myriad issues not the least of those being climate change adaptation; further, national interest proves contingent upon the inherently subjective beliefs of the individuals holding positions of legitimate authority in government. While diplomacy can play a role in setting conditions for action and shaping subsequent national behaviour in the form of policy, the development of "new technologies, business practices, and incentives for transformation—alter the realm of what is politically possible," which indicates that it might not be necessary for all states to act in sync [41, p. 1]. Developments in these areas depend upon innovation by private industry that would create new capabilities and opportunities for states to combat the effects of climate change, while also increasing the persuasiveness of negotiating positions used in diplomatic engagements due to the enhanced effectiveness of the new technologies or business practices.

If the present states of technology and business prove the limiting factors to counteracting the effects of climate change effectively and not the terms of international agreements achieved through diplomatic efforts, then the exercise of persuasive conditioned power more closely resembles an exercise in futility and therefore would not meet the effectiveness criteria for constituting an exercise of any type or form of power. The premise that diplomacy does not offer a viable pathway to mitigating the effects of climate change implies that at present any exercise of national power directed against climate change inevitably will fail. Despite what might amount to much ado about nothing, the process of achieving the Paris Agreement did increase global awareness and indicate solidarity, even if fleeting, among the community of states that the changing climate poses a significant threat to global stability. The consequences of the United States abandonment of the Agreement and the climate change threat do not endanger the possibility of reaching future agreements that incorporate new technologies and business processes coupled with enforcement and accountability measures, but rather threaten to neuter the ability of the United States to influence any such future agreements. Withdrawal from the Agreement will cost the United States in terms of prestige, which will correspond to its ability to wield diplomatic national power, understood as persuasive conditioned power. Loss of prestige for any state therefore proves counter to its national interest because of the corresponding opportunity loss to exercise its power. The chief negotiator from the United States delegation to COP21 affirms the cost and outlines the loss of power that the transience of interest caused: "At the senior, political level, the United States is simply not participating. And, in the wake of both President Trump's announced intention to withdraw from the Paris Agreement and his administration's overt antipathy to any policy meant to address climate change, the United States has lost much of the leverage and credibility it built up during the Obama years" [27, p. 8]. The shift in ideologies and corresponding actions precipitated by the change in presidential administrations undermined the leadership role on climate diplomacy that the United States had earned by developing relationships with key partners including rivals like China, leading working groups like the Major Economies Forum on Energy and Climate (MEF) that met semi-annually to advance negotiations, and convening key players when necessary to resolve significant issues—all actions that were reinforced by an administration that aligned its strategic messaging and actions on climate change [9, p. 18, 27, p. 8]. The new administration defined its approach to climate change as the antithesis to the previous one and set out to undo progress and disengage the United States officially from the Paris Agreement and unofficially from policies that recognize climate change as scientifically legitimate, seemingly only for the sake of appeasing the anti-science ideology present within the political party that facilitated the ascendency of the new administration. This shift reaffirms that dependence between identity, interest, and action, evidenced by the contrasts between beliefs and actions on climate change between two consecutive government administrations in the United States.

5.10 The Transience of National Identity in the United States

The national identity of the United States proves transient based on the public policies that a given presidential administration pursues. Role-taking and alter-casting therefore emanates from the dispositions of the political party from which a given president comes. Between the founding of the UNFCCC in 1992 and reaching the Paris Agreement in 2016, the United States experienced two non-consecutive administrations of eight years each that sought to implement policies that would mitigate the effects of climate change [11, 25, pp. 415–416]. By pursuing these policies, the United States took on the role of champion for climate change action within the international community, along with other states that adopted similar policies and contributed to the process culminating in the Paris Agreement. As a consequence of taking on this role for itself, the United States cast other states that did not pursue similar policies as others within the international community; similarly, individuals and groups within the United States that disagreed with pro-climate change adaptation policy became cast as others. The act of role-taking produces the reciprocal effect of alter-casting. For policy issues like climate change adaptation, identity formation follows political party lines and becomes an ingrained aspect of party identity such that the party, as an organization, becomes less likely to tolerate moderate positions, which creates conditions that favour radicalized manifestations of once reasonable convictions. When an unabashed anti-climate change adaptation administration won election in 2016 it abdicated the Paris Agreement to spite its predecessor more so than based on national interest, despite declaring it as a key part of pursuing an America First policy [29]. In addition to losing its standing as international climate change champion, the United States lost time that proves critical for research, development, and implementation of technological and social solutions that would have been implemented from 2017 to 2021 to combat climate change. The lost time and unimplemented policies and technologies result from a change in policy that arises from a transient national identity tied to political parties and the pendulum on which they and the American voters swing quadrennially.

The abdication by the United States of its role as leader or champion for global climate change adaptation not only costs its own prestige and corresponding ability to exercise persuasive conditioned power but also causes the international momentum to falter, even if briefly until another entity fills the vacuum. Change of any kind threatens stability, consensus, and any action that presupposes either, so creating any political vacuum proves at least potentially counterproductive if not entirely damaging due to the difficulty associated with change. Vacuums persist for a very little while and once filled, regardless by which entity, the United States and the perception of its role as global champion will have been greatly diminished by its own hand. In this case, the interest of the United States and global interests suffer.

5.11 Withdrawal and the Damage Done

The United States withdrawal from the Paris Agreement highlights the difference between exercising power within a group as part of it, as opposed to exercising power against a group from outside of it. Remaining with an organized group of equals unites their purposes in a way that reduces the inherent reluctance to collaborate with others [23]. While each retains its separate identity and corresponding interest, contributing to the organization engenders positive relationships that prove critical when attempting to solve problems that surpass the capacity of any individual member, or in the case of climate change adaptation any combination of members less than all. Participating in organized efforts facilitates information sharing, setting of priorities, resource pooling, and orchestrating a coordinated, mutually reinforcing strategy intended to benefit the collective. By contributing to the passage of the Paris Agreement, the United States used its position to seek an arrangement that would suit its interest first and because during that time it had aligned its interest with the global interest no discrepancy appeared between them. Exercising power as a constituent proves more effective than as an unaffiliated other lamenting the organization from beyond its membership. Even if the United States altered the way in which it defines its national interest, it proves more effective to remain as a protesting member than withdraw; the virtue of the loyal opposition can only come from criticizing an organization within its ranks. The decision by the United States to leave the Paris Agreement therefore constitutes self-othering; having determined that participating in the Agreement does not serve its national interest, the United States departed the group and ascribed for itself the role of antagonist while casting the states party to the Agreement as conforming to external pressures rather than internal determination [42, p. 329]. One of the perks of membership is the ability to drive progress through words and deeds, and those that have banished themselves can lay no claim to such opportunities. Juxtaposed with the external antagonist, membership of any kind in an organization offers a greater opportunity to exercise power—especially for the social forms of power and authority, which account for all of types and combinations thereof save condign power [24, p. 86].

The power of disbelief manifests as a reluctance to accept the arguments presented during an attempt to exercise persuasive conditioned power. By withstanding the attempt, the beliefs of the intended power subject—including constitutive elements of identity and interest—remain unchanged by the information offered. For the United States, seeking to withdraw from the Paris Agreement represents disbelief, or at least reluctance to accept, that "science has proved the reality and threats of climate change, and may have proved human complicity in contributing factors, it has not proved their exclusivity, much less the causality to a reversal from reducing the carbon footprint of homo sapiens to the levels of a millennium ago" [4, p. 9]. This reluctance to accept scientific fact, especially inconvenient ones, justifies the political actions of the administration, which in this case accounts for the United States withdrawal from the Paris Agreement. Rejecting an exercise of persuasive conditioned power depends upon not only the assessment of the ideas presented, but also how accepting those ideas might compel the power subject to alter his or her life to what would be newly adopted ideals. Given that "human behaviour, especially if it carries substantial commercial and economic benefits in the short term, is highly unlikely to change quickly and substantially on a voluntary basis," the argument underlying an attempt at persuasion must surpass the ability of the power subject to resist through either doubt or anxiety stemming from the transition from one belief set to another [4, p. 6].

The introduction of doubt weakens resolve at both the collective and individual levels and proves a formidable adversary for persuasive conditioned power. Sustained down turns to scepticism having risen from a lack of trust and confidence in either the message, messaging, or messenger; that is, what is being presented, the way in which its being presented, or who is presenting it. Doubt in any of these three areas reduces the likelihood that the information will be accepted. When individuals do not believe in an issue, it reduces the resolve of the collective to act on that issue; this is the inverse effect of bimodal symmetry. Just as belief by individuals comprising a collective in its purposes increases the effectiveness of the entity, disbelief or doubt held by constituents decreases the ability of the collective to act.

5.12 Conclusion

Truth exists independently from the opinion that one has of it. In terms of climate change and its consequences, this proves gravely significant due to the requirement for collective action to mitigate its effects. The consequences of climate change will continue to manifest despite the degree of belief in the causes of the phenomena. The problem of collective action relates to the challenges associated with competent authority, specifically in its dependence on the belief of the power subject that invests competent authority into the power holder. Belief requires the subject to recognize supremacy in the knowledge, skill or ability held by the holder, in this case that of knowledge as a form of intellectual property tantamount to wisdom upheld by science and therefore beyond reproach, which serves as the source of power from which the competent authority derives its ability to influence any potential power subject. The challenge for global climate change adaptation therefore mirrors the problem associated with exercising persuasive conditioned power at the individual level: "global consciousness of global problems and the realization of a need for global solutions exists no more than support for global government. It is fair to conclude that, in an age where not even the ideology of nation states can be overcome, the time for this type of solution has not yet arrived" [4, p. 6]. While collective global action on climate change remains challenging due at least in part to disparate national interests, many states have incorporated climate change adaptation as central policy goals and remain committed to the goals of the Paris Agreement [34].

Climate change adaptation will remain a significant challenge for the United States until it can adapt its national identity and interest to account for the threat posed by climate change and implement a combination of technological and social solutions to improve business process within the framework of a viable global strategy that allocates resources and sets priorities to mitigate the consequences of climate change. Forward-looking statements from the 2013 President's Climate Action Plan offer a glimpse of the beliefs that led to the Paris Agreement:

> Climate change represents one of our greatest challenges of our time, but it is a challenge uniquely suited to America's strengths. Our scientists will design new fuels, and our farmers will grow them. Our engineers to devise new sources of energy, our workers will build them, and our businesses will sell them. All of us will need to do our part. If we embrace this challenge, we will not just create new jobs and new industries and keep America on the cutting edge; we will save lives, protect and preserve our treasured natural resources, cities, and coastlines for future generations [11, p. 5].

The administration that followed did not share the belief in the necessity of dedicating these resources to combat the effects of climate change, and its election indicates that the same holds true for a significant number of United States citizens. The corresponding shift from defining interest globally in terms of the collective towards singularly in terms of the individual amounts to an exchange of "we" for "I" that propagates otherness and division, which stand in contrast to global stability. The divorce of the United States from global interest proves detrimental to both the collective and the individual.

The model introduced within this chapter illustrates the United States exercise of power in advocating for the Paris Agreement of 2015 and those leading up to its passage by the UNFCCC. Two lines of effort appear, with one drawing on personality and the other organization as sources of power. Both sources of power enable conditioned power as the instrument through which the United States acts to persuade the other parties to the several intersubjective social power relations comprising the episode—for the advancement of the agenda that became the Paris Agreement, these relations include all other states via their diplomatic representation to the Conferences of the Parties and the interactions with the United States representatives. The model shows how the United States exercised persuasive conditioned power as national power, which transcends each of the four domains. While negotiations transpire primarily through the diplomatic and information domains, the consequences of accepting or rejecting the Agreement manifest within the economic and military domains in the forms of nationally determined contributions and the impacts of policies required to achieve the goals and potential exacerbation of geopolitical stressors leading to increased conflict, respectively. This phenomenon suggests that an exercise of national power transcends the four domains in such a way that each exercise contains elements of all domains, even if defined beyond the four that this book examines. The model also shows how the United States exercised persuasive conditioned power to justify its withdrawal from the Paris Agreement. While the exercise transpires in the same way in terms of sources, instrument, and forms, the intersubjective social power relation transpires between the government of the United States and its people. Within this relation, the role of identity and interest appears as the decisive factor. The national interest of the United States reflects the identity of its elected administration, which represents general sentiment within the United States electorate. National identity therefore derives its character from the beliefs held by constituents; changes in belief within the electorate correlate to changes in national identity, interest, and exercises of power that serve both. Countervailing exercises of power like those that set out to advance what became the Paris Agreement and then abandon it suggest the presence of cognitive dissonance within the collective American psyche.

References

1. American Federation of Labor and Congress of Industrial Organizations (1998) Executive Council Statement on the Kyoto Protocol. Washington, DC; January 30, 1998. https://aflcio.org/about/leadership/statements/kyoto-protocol. Accessed 7 Jan 2022
2. Baumol W, Oates W (1988) The theory of environmental policy, 2nd edn. Cambridge University Press, New York
3. Brookes, T and Nuthall, T (2009) What did the Copenhagen climate summit achieve? The European Climate Foundation; BBC News. http://news.bbc.co.uk/2/hi/science/nature/8424522.stm. Accessed 7 Jan 2022
4. Catyas, JD (2018) The COP21 Negotiations one step forward, two steps back. consilience, No 19, pp 1–16. Columbia University Stable. https://www.jstor.org/stable/26427709. Accessed 7 Jan 2022
5. Chakraborty B (2017) Paris Agreement on climate change: US withdraws as Trump calls it 'unfair.' Fox News; June 1, 2017. https://www.foxnews.com/politics/paris-agreement-on-climate-change-us-withdraws-as-trump-calls-it-unfair. Accessed 7 Jan 2022
6. Dasgupta D, Espagne E, Hourcade J-C, Minzer I, Nafo S, Perissin-Fabert B, Robins N, Sirkis A (2016) Did the Paris Agreement plant the seeds of a climate consistent international financial regime? Fondazione Eni Enrico Mattei (FEEM). http://www.jstor.com/stable/resrep15111. Accessed 3 Dec 2020
7. De Bruin B (2008) On glazer and rubinstein on persuasion. In: Apt KR, Van Rooij R (eds) New perspectives on games and interaction. Amsterdam University Press. https://www.jstor.org/stable/j.ctt46mwfz.10. Accessed 2 Dec 2020
8. Derwent H, Błachowicz A, Hügel J, Fernando Blanco L, Low Yu Xing M, Moreira Franco N (2016) Analysis of the Paris Agreement Report; Paris Agreement: how it happened and what next: with a special focus on the role of non-state actors. Climate Strategies. http://www.jstor.com/stable/resrep15582.4. Accessed 2 Dec 2020

9. Engberg-Pedersen L (2011) Institutional issues in international climate change cooperation. In: Climate change negotiations and their implications for international development cooperation. Danish Institute for International Studies. http://www.jstor.com/stable/resrep15638.6. Accessed 3 Dec 2020
10. Ewing J (2012) From Kyoto to Durban: the fits and starts of global climate change negotiations. S. Rajaratnam School of International Studies. https://www.jstor.org/stable/resrep26795. Accessed 2 Dec 2020
11. Executive Office of the United States President (2013) President's climate action plan. Washington, DC. https://obamawhitehouse.archives.gov/sites/default/files/image/president27sclimateactionplan.pdf. Accessed 7 Jan 2022
12. Friedman L (2019) What is the green new deal? A climate proposal, explained. New York Times; February 21, 2019. https://www.nytimes.com/2019/02/21/climate/green-new-deal-questions-answers.html. Accessed 3 Dec 2020
13. Funk C, Hefferon M (2019) US public views on climate and energy. Pew Research Center. https://www.pewresearch.org/science/2019/11/25/u-s-public-views-on-climate-and-energy/. Accessed 3 Dec 2020
14. Galbraith J (1983) The anatomy of power. Houghton Mifflin Company, Boston
15. Goulder L, Nadreau B (2002) International approaches to reducing greenhouse gas emissions. In Schneider StH, Rosencranz A, Niles JO (eds) Climate change policy, pp 115–150. https://web.stanford.edu/~goulder/Papers/Published%20Papers/Intl%20Approaches%20for%20GHG%20Redux%20-%20Goulder-Nadreau.pdf. Accessed 7 Jan 2022
16. Hahn R, Stavins R (1999) What has Kyoto wrought? The real architecture of international tradable permit markets. Discussion Paper 99-30, Resources for the Future, Washington, DC. March, 1999. https://www.rff.org/publications/working-papers/what-has-kyoto-wrought-the-real-architecture-of-international-tradable-permit-markets/. Accessed 7 Jan 2022
17. Kenton W (2019) Cap and trade definition. https://www.investopedia.com/terms/c/cap-and-trade.asp. Accessed 7 Jan 2022
18. Kilcullen D (2014) Out of the mountains. Oxford University Press, New York
19. Memmott M (2009) Obama in Copenhagen; climate talks in disarray; urges 'action over inaction'. National Public Radio; December 18, 2009. https://www.npr.org/sections/thetwo-way/2009/12/obama_in_copenhagen_climate_ch.html. Accessed 7 Jan 2022
20. Nakamura K, Epstein S (2007) Diplomacy for the 21st century: transformational diplomacy. Congressional Research Service, Washington DC, August 23, 2007. http://www.fas.org/sgp/crs/row/index.html. Accessed 10 Mar 2021
21. Obama B (2009) Address to the United Nations General Assembly. New York, NY, September 23, 2009. https://ecf.org.il/media_items/1147. Accessed 7 Jan 2022
22. Obama B (2015) Statement by the president on the Paris climate agreement, December 12, 2015. https://obamawhitehouse.archives.gov/the-press-office/2015/12/12/statement-president-paris-climate-agreement. Accessed 19 Apr 2022
23. Powell J, Menendian S (2020) The problem of othering: towards inclusiveness and belonging. Othering and belonging. https://www.otheringandbelonging.org/the-problem-of-othering/. Accessed 3 Dec 2020
24. Rose S (2020) Elite incentives and power dynamics in fragile states. PRISM 8(4):84–99. Institute for National Strategic Security, National Defense University. https://www.jstor.org/stable/10.2307/26918236. Accessed 3 Dec 2020
25. Royden A (2002) U.S. climate change policy under president Clinton:, 32 Golden Gate U. L. Rev. http://digitalcommons.law.ggu.edu/ggulrev/vol32/iss4/3. Accessed 3 Dec 2020
26. Stavins R (2001) Experience with market-based environmental policy instruments. Discussion paper 01–58, Resources for the Future, Washington, DC. November 2001. https://media.rff.org/archive/files/sharepoint/WorkImages/Download/RFF-DP-01-58.pdf. Accessed 7 Jan 2022
27. Stern T (2018) The Paris Agreement and its future. The Brookings Institution, Washington, DC. https://www.brookings.edu/wp-content/uploads/2018/10/The-Paris-Agreement-and-Its-Future-Todd-Stern-October-2018.pdf. Accessed 7 Jan 2022

28. Tamura K, Suzuki M, Yoshino M (2016) Raising ambition towards the 1.5°C/2°C goal: ratcheting-up mechanism under the Paris Agreement. Institute for Global Environmental Strategies. http://www.jstor.com/stable/resrep02905. Accessed 2 Dec 2020
29. Trump D (2017) Statement by President Trump on the Paris Climate Accord. June 1, 2017. https://www.whitehouse.gov/briefings-statements/statement-president-trump-paris-climate-accord/. Accessed 3 Dec 2020
30. United Nations (1992) United Nations framework convention on climate change. https://unfccc.int/files/essential_background/background_publications_htmlpdf/application/pdf/conveng.pdf. Accessed 7 Jan 2022
31. United Nations (1998) Kyoto protocol to the United Nations framework convention on climate change. https://unfccc.int/resource/docs/convkp/kpeng.pdf. Accessed 7 Jan 2022
32. United Nations (2015) The Paris Agreement. https://unfccc.int/sites/default/files/english_paris_agreement.pdf. Accessed 7 Jan 2022
33. United Nations (2019) Paris Agreement status of ratification. https://unfccc.int/process/the-paris-agreement/status-of-ratification. Accessed 7 Jan 2022
34. United Nations Climate Change (2020) Status of Paris Agreement. https://unfccc.int/process/the-paris-agreement/status-of-ratification. Accessed 3 Dec 2020
35. United Nations Framework Convention on Climate Change (2020) Gender composition. New York; October 7, 2020. https://unfccc.int/documents/254399. Accessed 2 Dec 2020
36. United States Department of Defense (2010) Quadrennial defense review. Washington, DC. https://unfccc.int/resource/docs/convkp/kpeng.pdf. Accessed 7 Jan 2022
37. United States Department of Defense (2014) Quadrennial defense review. Washington, DC. https://www.acq.osd.mil/ncbdp/docs/2014_Quadrennial_Defense_Review.pdf. Accessed 14 Feb 2025
38. United States House of Representatives (2019) House resolution 109. https://www.congress.gov/bill/116th-congress/house-resolution/109/text. Accessed 3 Dec 2020
39. United States Marine Corps (1997) Marine corps doctrinal publication 1 Warfighting. United States Department of the Navy, Headquarters Marine Corps. Washington, DC. https://www.marines.mil/portals/1/publications/mcdp%201%20warfighting.pdf. Accessed 2 July 2021
40. United States Senate (1997) Senate resolution 98. https://www.congress.gov/bill/105th-congress/senate-resolution/98/text. https://www.federalregister.gov/documents/2017/01/19/2017-01369/continuation-of-the-national-emergency-with-respect-to-ukraine. Accessed 7 Jan 2022
41. Victor D, Jones B (2018) Undiplomatic action: a practical guide to the new politics and geopolitics of climate change. The Brookings Institution, Washington DC. https://www.brookings.edu/wp-content/uploads/2018/02/esci_201802_undiplomatic_action.pdf. Accessed 7 Jan 2022
42. Wendt A (2017) Social theory of international politics. Cambridge University Press, Cambridge
43. Wrong D (1979) Power: its forms, bases, and uses. Transaction Publishers, London and New Brunswick

Chapter 6
Belief, Identity, Interest, and Action: The Role of Narrative and Image in the January 6 Attack on the US Capitol

> Man is not a rational animal; he is a rationalizing animal.
>
> Robert Heinlein [11]

> Limited in his nature, infinite in his desires, man is a fallen god who remembers the heavens.
>
> Alphonse de Lamartine [6]

On January 6, 2021, then-President Donald Trump addressed a crowd at the White House ellipse that he had gathered to protest against the certification of the 2020 general election, which he claimed had been rigged, fraudulent, and corrupted. Trump had summoned the group to Washington, DC over the preceding weeks, motivating their action by claiming that he had won the general election in November 2020 and that massive election fraud caused his defeat. He created a narrative in which he was the victim of a massive conspiracy perpetrated against him and the will of the people that would destroy the foundation of the United States by undermining the authority of the US Constitution and the voice of the people. Trump argued that not only was he being persecuted, but also were his supporters. The words of the sitting President resonated with his most loyal supporters and those who saw an opportunity in his vitriol to enact their dark fantasies of violent revolution against the American system of government. The events that transpired following Trump's speech have been described as an insurrection, attack, riot, and assault against the institutions of the United States government as outlined in the Constitution and put into action by the United States Congress in the Capitol in response to the votes represented by the electoral college that would remove Trump from office two weeks from January 6 [8].

K. L. Wedekind, *Constructing Power: A Critical and Constructivist Theory Using the American Case*, Advanced Sciences and Technologies for Security Applications,
https://doi.org/10.1007/978-3-031-84893-3_6

This case study is about power. It is not about justice, morality, ethics, or reason; it is not about religion or faith. This case study examines how Trump wielded his power to inspire the attack on the US Capitol to disrupt certifying the votes of the electoral college and why his narrative resonated so profoundly with those who participated in the attack. The case study focuses on the role that beliefs, identity, and interest play in determining how one exercises power to understand more clearly why January 6 happened. Trump exercised power over his followers and succeeded in disrupting Congress in the act of certifying his defeat but failed to stop it and achieve his desired goal; it also provided an opportunity for his supporters to express their rage against what they perceived as a corrupt system working against their beliefs, identity, and interest. This episode provides clear opportunities for analysing role-taking and altercasting; a single source of power working across each instrument and form of power; and a juxtaposition of Trump's intentionality and effectiveness as the power holder and the extensiveness, intensiveness, and comprehensiveness of his actions upon his followers. By focusing on the power relation between Trump and his followers, this case study seeks to understand how the exercise of power culminating in the January 6 attack on the US Capitol transpired and considers whether it succeeded or failed.

Figure 6.1 outlines the connections between the source, instruments, and forms of power wielded to drive the J6 attack on the US Capitol. The sole source of power in this episode emanated from President Trump, then as the outgoing US President. The primary line of effort flowed through conditioned power to manipulation supported by condign power and force, compensatory power and persuasion, and conditioned power and authority, respectively. The secondary line of effort uses condign power and force in the form of psychological and physical violence with the former continuously appearing in speeches, social media posts, and litigation while the latter manifests in the January 6 attack. The third line of effort uses compensatory power to imply a reward for accepting one side of a persuasive argument. The fourth line of effort exists as an implication that all acts undertaken by a sitting President constitute an official act by virtue of the position and carry with them the authority of the President. The combination of multiple, mutually reinforcing lines of effort produced high degrees of effectiveness and intentionality as well as intensiveness and comprehensiveness among the power subjects.

6.1 The Savior's Introduction, Persecution, and Resurrection

In 2015, Trump introduced himself as a candidate for President of the United States by descending upon a golden escalator in Trump Tower, a New York City building that he had named after himself [13]. This role-taking act birthed the movement that would propel Trump to victory in the 2016 US general election. The moment itself contained clear implications that Trump would present himself to potential supporters as a saviour. The metaphors of the scene reveal their meaning. In descending the

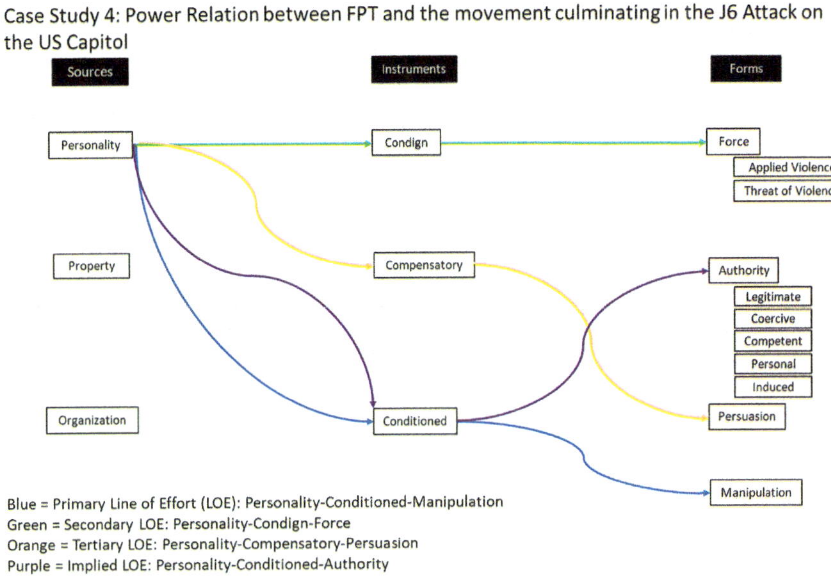

Fig. 6.1 Power relations between President Trump and the crowd that gathered on January 6 in Washington, DC (Author's own)

escalator, Trump physically moves from an elevated to a lowered position, which represents an act of sacrifice as he relinquishes the safety of his elevated position in society as a billionaire celebrity businessman to enter the fray of the political contest. The crowd into which Trump descended wore campaign merchandise emblazoned with "Make America Great Again," the slogan that created the necessity for the saviour by implying that the United States had once been a great nation and was no longer one but could be again with Trump leading it [13]. The complementary role-taking acts of descending the escalator and manufacturing the need for a saviour created the political narrative that fuelled Trump's campaign and 2016 electoral victory.

Political candidates and their staffs design campaign events to create an image of the candidate intended to resonate with the targeted voting bloc and motivate their support. A narrative accompanies the visuals to inspire beliefs where they do not align with those presented by the candidate and to enflame them where they do. The image and narrative presented by President Trump during his first campaign reflected this deliberateness and design though the content and success differed significantly from any other modern political candidate. The image of a successful, celebrity businessman descending an escalator in a building that he owns to a sign-waving crowd implicitly signals the arrival of the saviour. The narrative reinforces the image as the businessman willingly sacrifices his world of luxury for the political campaign, which is intended to portray the man as selfless and driven by a mission to right the supposed wrongs in society that have harmed the American people. These actions mirror the first stage of the monomyth identified by Campbell [4] as the underlying master narrative in myths and initiation rites that follow three distinct stages of initiation, departure, and return (p. 58). By descending the escalator, Trump entered the fray of the everyday world and initiated his journey to "Make America Great Again." The narrative claiming that the US has lost what once made it great combined with the image of the selfless saviour create the problem and provide the solution, respectively.

These actions constitute the initial role-taking and alter-casting that have defined Trump as politician. The departure from safety into the unknown takes on the role of the hero or the saviour on a mission while simultaneously casting all alternatives as obstacles or impediments to accomplishing the mission. This creates a false dichotomy that reflects the idea that there are only two options available—good or evil—and everyone must choose whether to align themselves with one of these forces. The narrative presents no other options so that the power subjects feel compelled to choose. Power operates here through the personality of the power holder and his ability to wield conditioned power through persuasion. The image and narrative present an argument while the mechanisms used to present it also seek to deny and discredit alternatives by appealing to preexisting beliefs and the predisposition to distrust the legitimate authority of the government. This is the beginning of cognitive dissonance in the argument. In the United States, government famously exists "by the people, of the people, for the people" through the institutions created by the US Constitution [14]. To attack the legitimate authority of the government is to attack the foundation upon which it rests, which is the Constitution and by extension the collective, sovereign United States citizenry. Given the logical fallacy imbedded within the circular argument, it becomes necessary to create a sinister entity to blame that corrupts the will of the people as expressed through government.

The so-called "deep-state" emerges within the Trump narrative as the scapegoat that perverts the American system of government and usurps the power of the American people through "rigged" elections, massive fraud, and corruption [25]. Framed in this way, the narrative separates perception and reality. The power subjects receive affirmation for their grievances which generates a positive response to the message and the messenger. The reality of the situation—that the government in the United States operates according to the rule of law that derives its authority from the US Constitution and the laws established through the processes it mandates with oversight through checks and balances from different branches of the federal government and the legitimate authority of state governments—becomes irrelevant because the narrative presents the government as a monolithic organization possessing its own singular agency rather than a composition of individuals who come from the same collective body, the US citizenry, as do those who seek to disrupt or destroy its processes and mandate [20]. An organization, even a governmental body, is simply a group of people working in coordination with one another according to a shared vision or framework. Organizations with high degrees of bimodal symmetry appear to take on their own agency. The individuals comprising the organization reflect the shared qualities and characteristics that bind them to the other participants and present an image to outsiders; that is, organizational culture produces its own role-taking and alter-casting through an assimilation process that mirrors the role-taking and alter-casting that transpires in political campaigns. The power subjects in both cases incorporate their beliefs into their identities. Higher intensiveness and comprehensiveness of belief increases the impact on the identity of the individual which consequently defines how the individual determines their interest and how one exercises power to satisfy it.

The Trump image and narrative presents multiple false dichotomies of good and evil, hero and villain, and with me or against me. It affords the power subjects with the appearance of simple decisions but is begging the question by presupposing what it seeks to establish. The central element of this circular reasoning is its insistence that the United States was once a great nation, is now in decline, and can only be saved by one person who will restore the conditions that once made it great. Trump inserts himself into this narrative as the saviour consistently; he is the one constant in the equation and treats the other parts as variables, inserting whichever antagonist has gained enough public attention to resonate with the intended audience. Persecution of Trump as saviour appears as the principle, recurring theme. It is an effective rhetorical construct even if lacking in elegance because it resonates with Christian nationalists, a preexisting movement in the United States that works along a parallel path to subvert American democracy in favour of authoritarian theocracy [22]. The alignment of goals between Trump and the Christian nationalists supporting him justify interpreting his role in present day events in a way that avoids a crisis of faith among the faithful [24]. The Christian nationalist necessity of a saviour figure reflects the principles in the dogma of the Christian faith wherein Jesus Christ, as one part of the Trinity and therefore God, becomes man to sacrifice himself to save humanity from their sins and eternal damnation. While such a comparison would constitute blasphemy, the subtext and implied comparison between Trump and Jesus Christ remains. Christian Nationalist leaders have attempted to avoid this overt comparison by insisting that Trump's ascension more closely reflects that of Cyrus, the Persian King identified in the Old Testament of the Bible as an instrument working to implement God's will [3, 9, p. 15].

The methodological alignment of narratives between Trump and Christian nationalists proves mutually beneficial. Trump desires the role of the saviour and Christian nationalists need one. Both target every out-group as a potential threat and rely upon true believers to recruit and convert more followers while increasing the intensiveness of their belief.

6.2 The Wrath of the Savior

On November 3, 2020, then President Trump lost the election to former Vice President Joseph Biden. Between his defeat on election day and the inauguration of his rival, Trump initiated a crusade to challenge the election results and remain in office. During the weeks from election day to January 6, Trump worked through multiple lines of effort to achieve his desired end state—to exercise his power to satisfy his beliefs about his own identity and interest—by remaining President despite losing the 2020 election. Trump (1) repeated false claims of widespread election fraud as the cause of his defeat through social media posts; (2) filed over 60 lawsuits against individuals and organizations responsible for administering elections in the states and localities where he had lost; and (3) conspired with a surrogate legal team to use demonstrably illegal means and justifications to overturn the electoral process [8]. These actions culminated on January 6 as an attack on the US Capitol as Congress attempted to certify the votes of the electoral college. Outgoing President Trump orchestrated and inspired the attack through his words and actions, including a speech at the White House ellipse to the group that had gathered on the day of the attack. During the speech, Trump revealed to the crowd that the Vice President had upheld his constitutional duty by defying Trump's call for him to intervene in the electoral count rather than overseeing it [25]. Trump insisted the opposite as truthful and continued his manipulation campaign by presenting an alternative set of facts to a sympathetic audience that believed more in the messenger himself than the objective facts that falsified the manufactured narrative. Once Trump's speech ended, the group marched from the White House ellipse to the US Capitol and began an assault into the chambers where the mob disrupted and delayed the certification of the electoral process, assaulted over 140 law enforcement officers, damaged property, and defiled the sanctuary of American democracy that represents "we the people" within the United States government [8].

Personality, conditioned power, and manipulation serve as the operative elements at work through the primary line of effort that Trump used to generate support for his attempt to remain as President. This effort manifests as communications reliant upon a manufactured narrative repeated as frequently as possible to shape the beliefs, identities, and interests of the individuals comprising the intended audience while simultaneously increasing the intensiveness and comprehensiveness of belief within this same community. The combination of press conferences, including the Four Seasons Total Landscaping event on November 7, 2020, and social media posts that presented false information as factual despite no evidence supporting the claims created an information vortex swirling around the intended audience where objective facts became a secondary consideration to their perception of a grievance against the preeminent figure in their perceived worldview and by extension themselves. These actions constitute an exercise of conditioned power through manipulation by leveraging the illusory truth factor that occurs when an individual begins to reshape their beliefs based on the number of repetitions of information as opposed to its credibility [17]. Facts became irrelevant because the audience cared more about the central figure of the narrative than whether he told the truth or remained duty bound by his oath to the US Constitution. Trump manipulated his followers and those who lacked enough understanding of a complex system of government by sowing the seeds of doubt and distrust through the replacement of reality with perception; that is, if one can convince another that their perception matters more than reality, the latter becomes irrelevant. In this environment, the individual prioritizes belief over fact making belief even more susceptible to manipulation by removing the ties that bind it to its ontological and epistemological origins. Belief, inherently subjective, becomes dependent upon the information the individual consumes in a way that redefines identity through role-taking and alter-casting against the new other as antagonist. Interest follows suit and the exercise of power becomes an attempt to satisfy oneself.

6.3 Imperial Christianity Marches Onward

The narrative of a fraudulent election that Trump used between November 3 and January 6 did not contain any content specific to the Christian faith that would reasonably cause the many denominations of Christianity in the US to congregate as a unified body in Washington, DC on January 6. It did contain key elements of the narrative that Christian nationalists use to justify their beliefs, identity, interest, and the ways in which they use power to implement a vision of the United States based on a manufactured narrative of an ahistorical past [9, p. 90, 19, 24]. Christian nationalism.

> is an identity built on the claim that America was founded as a Christian nation, that it was based on Judeo-Christian principles, and, most importantly, that we have strayed from that foundation, our godly roots. Christian Nationalism employs the language of return in an effort to claim to be the true heirs of the American experiment and, more importantly, the American Identity. In their retelling, Christian Nationalists are the true Americans, everyone else is an interloper [19].

The individuals that organized and led the preparations for January 6 used language that resonated with Christian nationalism groups because their beliefs predispose them to believing demonstrably false ideas that contradict the US Constitution and justifies the use of force, as both physical and psychological violence, to seize legitimate authority as a means of implementing its world view. Long-standing Christian nationalist arguments interpret the religious freedom outlined in the Establishment Clause, Article IV Clause 3, elsewhere in the US Constitution, and supporting case law protecting religious freedom as persecution of their beliefs by denying their right to use the legitimate authority of government to implement religious ideology through the law. The Christian nationalist perspective on religious freedom proves illogical as it demands that the law reflect their religious beliefs—that one faith, the Christian nationalist version of Christianity, should influence policy and any resistance to its implementation constitutes persecution of their values and beliefs [18, p. 5]. This resistance is the inherent reaction within the rule of law that prevents the establishment of any religion though Christian nationalists believe it is persecution [24].

The Christian nationalist belief that all perceived deficiencies and problems in the United States have been caused by an abandonment of Judeo-Christian values, otherwise called traditional values, ensures that the United States will always appear in decline [23]. The perpetual decline of the United States creates a demand for a saviour who can bring about a return to traditional values and thereby restore the US to what Christian nationalists consider its rightful origins—to "Make America Great

Again." This desire is predicated on a desire for a fictional past based in alternative histories deeply ingrained as truth that forms the foundation of the Christian nationalist belief set [18, pp. 10–11]. The language of return serves as a *fait accompli* as there is no previous period in US history in which Christianity abided; likewise, it is a matter of opinion as to whether the United States has been great or if it has when that might have been. Christian nationalists manufacture an ahistorical narrative of past greatness to justify bringing about systems change for their desired and unprecedented future wherein the United States becomes a Christian nation as the expense of its Constitution and its protections for religious freedom [24, p. 128]. Freedom of religion can only exist so long as freedom from religion persists.

Christian nationalist beliefs require a saviour willing to act on their behalf to bring about the change that their identity and corresponding interest demand. By offering to Make America Great Again, Trump presented himself as this saviour by embracing the core tenets of their beliefs focused on using a notional past to justify present political change. Opponents to the political movement within the Republican Party have been disparaged as Republicans in Name Only, or RINOs, which itself is an appeal to belief and identity as a means of shaming to induce compliance. Those who disagree with Christian nationalist dogmas are decried as not real Christians who will suffer eternal damnation for the heresy of resisting and upholding the secular Constitution and its provisions for religious freedoms. The acts of identifying an individual as a MAGA Republican or RINO and as a real Christian or not both constitute role-taking and alter-casting that form the identities of individuals within the in-groups [15, p. 163]. Reinforced by continued iterations results in a closer alignment of MAGA Republicans and Christian nationalists with the eventuality that these beliefs and identities integrate completely in a way that eliminates the possibility of resistance; that is, the beliefs indelibly shape the identities of the individuals captured by either of these dogmas in a way that drives their interests and how they exercise power.

This principle manifested on January 6 with the attack on the US Capitol as demonstrated by the composition of the attackers and their own role-taking on that day illustrated by their flags, clothes, and actions. The group had become enraptured by the "Big Lie" of a stolen election through non-existent fraud and identified their plight as the persecuted faction within the US with Trump's similar claims of injustice and unfair treatment. The false narrative from Trump resonated with people who already felt persecuted and offered a release of their disenfranchisement and anger with a system that refused to allow them to dominate the others, who account for all who had been alter-cast as part of the out-group relative to MAGA and Christian nationalists. Belief in their persecution as both Christian nationalists and as Trump-supporting MAGA Republicans—two key elements of their respective individual identities—compelled a corresponding change in interest that inspired the group to attack the US Capitol [19, 24]. Belief in persecution because of their identity motivated the escalation in terms of interest and the exercise of power as force in the forms of physical and psychological violence.

6.4 Against All Enemies, Perceived and Believed

On January 6, 2021, Trump succeeded in mobilizing a civilian mob to assault the US Capitol but failed to achieve his goal of disrupting the certification of his defeat. From the Oval Office as the then sitting President, Trump combined the elements of condign, compensatory, and conditioned power with force; coercive, competent, personal, and induced authority; persuasion; and manipulation to inspire an insurrection that originated from his personality as the single source of power. The property and organization of the US government and legitimate authority of his office at the time were the only elements of power that Trump did not use to support the actions of the civilian protesters turned attackers; neither did Trump use his legitimate authority as President to dedicate resources to defend against the attack but instead relied upon his personal authority to issue a statement to his mob to disperse, an order which the group followed. Personality therefore constitutes the sole source of power for the events of January 6.

Personality as a source of power emanates from physical and mental strength, agility, and endurance that fosters a willingness to act, which therefore bears its primary association with condign power as instrument and force as form, where condign power bestows punishments in the form of physical and psychological violence against the other (10, p. 6). Personality may also enable conditioned power as instrument and either persuasion or manipulation as form, as the success of these endeavours depends upon self-emanating conviction that resides within personality, which differs from identity. The former describes the capabilities of an individual and the various capacities thereof, while the latter pertains to the characteristics or traits that define personhood for the individual; that is, personality accounts for what an individual can do, and identity defines who the individual is. Identity therefore necessarily precedes personality.

From personality as the source of power flows the instruments of power used to drive the forms in which each would manifest. Conditioned power served as the primary instrument for bringing about the events of January 6 with manipulation as the form. Manipulation occurs when the power holder works to cultivate the subject's feeling of free choice through the presentation of information, which can be either true, false, or a combination of the two; even benign efforts by the power holder to encourage the subject to undertake or continue an activity in the subject's best interest as determined by his or her judgment can constitute manipulation if the subject acts unknowingly in accordance with the power holder's will. Calculation by the power holder and the ignorance of the power subject provide the two integral parts of manipulation as a form of power; when executed according to this definition, the holder denies the subject the ability to resist (Wrong 1979, p. 29). The exercise of power began in the aftermath of Trump's defeat as his claims of election fraud dominated his public remarks coupled with litigation filed by a separate legal team from the official White House General Counsel despite no evidence of widespread fraud and advice from his teams that he had lost the election. The absence of evidence and the fact that Trump's senior advisors confirmed his defeat in the electoral college and popular vote make Trump's claims of election fraud intended to undermine the results an attempted exercise of manipulation. By presenting a knowingly false narrative, Trump relies upon his authority to justify his argument rather than its merits to deceive the intended audience into believing him because of who he is and what he represents rather than the truth, which in this case emanates from the objective facts resulting from counting votes. The manipulation denies the audience, in this case the crowd that gathered on January 6 at the ellipse as power subjects, the choice that exists in the persuasive act of conditioned power.

Trump used manipulation to make the case for his own persecution in a way that resonates with his supporters, especially the Christian nationalist component, thereby shaping their perception and by extension their beliefs that call into question the validity of the election. Believing that the 2020 election had been marred by fraud and Trump had won despite no evidence served as justification for Trump's supporters to fund his efforts through donations and to participate in the January 6 rally turned attack. By claiming persecution, Trump took on the role of the victim and extends his circumstances to his followers as persecuted through him; he sought to bind his fate to his followers and by extension the United States as the persecuted victim to justify what would transpire on January 6 as "legitimate political discourse" [5]. By compelling belief in a fraudulent election and his own persecution, Trump appealed to the core elements of identity in his supporters as Americans, Christian nationalists, and MAGA Republicans who have an interest in protecting and defending the US against all enemies—foreign or domestic, but also real or perceived and believed—to exercise their power through planning and participating in January 6 or simply believing and repeating the lies about election fraud and continuing to doubt the validity of the electoral process. Trump used conditioned power and manipulation to shape beliefs that influence the identity and interest of his followers while justifying their exercise of power leading to January 6 and on that day.

The instrument of condign power in the form of force serves as the secondary line of effort through which January 6 occurred and perhaps the most acute and clear manifestation of the consequences of the effect of conditioned power and manipulation. Condign power "wins submission by the ability to impose an alternative to the preferences of the individual or group that is sufficiently unpleasant or painful so that these preferences are abandoned" (10, p. 4). Condign power maintains an association with negative corrective action or punishment, or the conference of a penalty upon another because of an unwanted or undesirable action "exercised by any form of adverse action or its threat, including fines, other property expropriation, verbal rebuke, and conspicuous condemnation by other individuals or the community" (10, p. 15). An exercise of condign power can occur through its actual application or through the threat of its use. In addition to physical force or harm, condign power includes effects upon psychological, mental, emotional, and spiritual faculties. Condign power therefore may target the physical body or metaphysical beliefs through actual application or merely the threat of its use to force the power subject to take the action or adopt the belief desired by the power holder.

The use of conditioned power and manipulation focused on influencing beliefs that would drive actions. These actions manifest through condign power and force as physical violence on January 6 and as psychological violence through implied and explicit threats against the United States and the American way of life. Psychological violence appears clearly on January 6 prior to the ellipse speech and during the attack at the Capitol. The gallows constructed near the Capitol steps represent premeditated intent to intimidate through psychological violence [8, pp. 63, 429, 529]. The calls and chants for the mob to use the gallows on many politicians—most notably the Vice President after the President condemned him on social media during the attack—further escalate this intent towards action (ibid). Some attackers wore shirts printed in advance to commemorate January 6 as "Civil War" even before the events had transpired; many carried flags reappropriated from the American Revolution, the Confederate States of America, QAnon, active National Socialist chapters reflecting a variety of political perspectives that include libertarianism, second amendment advocacy, Nazism, anti-government sentiments, conspiracy theories, and Christian nationalism [19, 21]. The successful breach of the Capitol allowed the forward line of attackers to reach within 40 feet of the Vice President [8]. While it may be unlikely that the attackers could have executed the Vice President outside the Capitol on the gallows constructed for January 6, the construction of the gallows constitutes an implicit threat. The social media posts driving its construction and advocating for its use in advance of January 6 coupled with the chants on that day to hang the Vice President constitute an explicit threat. While it is certain that the law enforcement officers protecting him on January 6 would have used expended every resource at their disposal to prevent that from happening, it is also certain that the attackers constructed the gallows as an instrument of condign power and force in the form of psychological violence [8, p. 63]. Given the clear intent of the attackers and the mission of the protective detail assigned to the Vice President, if the attackers had reached the Vice President, it is certain that the subsequent interaction between attackers and defenders would have resulted in the conversion of psychological violence and threats to physical violence and bloodshed. This did not happen, but even the spectre of this potential interaction that did not unfold haunts the mind of the defender while tantalizing the attacker in a way that perpetuates the use of psychological violence emanating from the events of January 6.

In every social relation, violence is always an option. Condign power and force resulted in physical violence on January 6. The crowd that gathered on that day to protest the results of the 2020 election based on calls to do so from the defeated but still sitting President arrived prepared to attack the US Capitol to disrupt the official proceeding of Congress that certified the votes of the electoral college [8, p. 5, 39]. After listening to a speech from President Trump at the ellipse, the crowd marched to the Capitol and attacked it. The attackers assaulted law enforcement officers, breached protective measures, and delayed the proceeding, but did not prevent it from transpiring later that evening.

Compensatory power operates as the tertiary instrument appearing in the form of persuasion. Persuasion always occurs as a social relation due to the necessity of interaction between power holder and subject. For an act of persuasion to occur, the holder must convince the subject to accept and comply willfully with the holder's desires. The subject weighs the arguments and rationale presented by the holder and chooses to do as the holder intends; the holder has convinced the subject of the proper course of action and induced compliance through a social relation. If persuasion is the tested acceptance of another's judgment, then authority is the untested acceptance of another's judgment; where the former relies on arguments, authority commands obedience and compliance from a position of power vested by an external source (Wrong 1979, p. 35) [7]. Trump makes his case to the judicial system and to the court of public opinion in an attempt to persuade both judges and the American people that his arguments have merit; in over 60 court cases, Trump lost all but one in resounding defeat of the merits of his legal cases including to 11 federal judges that Trump has appointed [8, p. 19, 26].

Trump's persuasion of the American public resulted in over $250 million in fundraising and the events of January 6 [8, p. 27]. Both investitures constitute compensation and indicate that the persuasive exercise emanated from the instrument of compensatory power. Where condign power punishes, compensatory power rewards. Compensatory power "wins submission by the offer of affirmative reward—by giving of something of value to the individual so submitting" [10, p. 5]. Compensatory power purchases submission, which requires the possession of property; this dynamic creates a dependency between property as source and compensation as instrument of power. The power holder rewards the power subject to ensure his or her submission in exchange for their compliance to the holder's will; in this sense, an individual or group induces submission through compensation that either generates a power relation or serves as positive reinforcement to sustain established and desired behaviour. Compensation may occur through the provision of access to resources that may include the physical, as in pecuniary reward or property, and the metaphysical, as in adulation or promotion. The purpose of the compensatory exercise mirrors that of the condign—to elicit submission of the recipient to the will of the giver. In exchange for the compensation received, the recipient agrees either tacitly or implicitly to comply with the requirements or demands levied upon him or her by the individual or organization that provides the compensation.

In this case and considered the totality of events leading to January 6, Trump exists as both power holder and subject due to the structural power shaping the circumstances he seeks to overcome. As the sitting but defeated President, he holds the legitimate authority of the office but is subject to the consequences of the electoral defeat. He pleads his case through traditional and social media to those sympathetic to his cause and susceptible to his persuasion while facing eviction, which makes him dependent upon the willingness of his supporters to fund his efforts through donations and to dedicate their own time and money to convene on January 6 to participate in the rally turned attack. The circumstances under which Trump pleads his case—as the outgoing, defeated incumbent seeking to retain his position—constitutes structural power that compels him to act simultaneously as power holder and subject. Trump presents his arguments through the instrument of conditioned power and manipulation, but these actions depend upon compensatory power and persuasion to drive the actions of the power subjects. Conditioned power and manipulation shape the perception of the power subjects and convince them that a problem exists while compensatory power and persuasion present Trump as the solution to the problem but ties his actions to a *quid pro quo* in which he can only act if sympathizers support him through donations and presence on January 6. Trump can become the saviour but only if his followers compensate him for it.

The influence of Trump's position as the sitting President while undertaking the actions that led to the January 6 attack serves as the fourth instrument at work. As President, his actions carry the weight of the office he occupied at the time and operate as conditioned power. Conditioned power manifests in the metaphysical and relies upon determinations made in the mind of the individual to act according to what one perceives as the preference of an agency external to the individual; that is, power subjects submit willingly based on a belief that to do so is the appropriate course of action. Galbraith explains the subjectivity of conditioned power as "neither those exercising it nor those subject to it need always be aware that it is being exerted. The acceptance of authority, the submission to the will of others, becomes the higher preference of those submitting" (1983, p. 24). It is impossible to separate the actions of the President from the legitimate authority of the Presidency as any action carries at least the implication of such. The President is a person, but the Presidency is an institution. Both have limits. Each reflects and shapes the other through actions, reactions, and counteractions that produce acceptance or resistance within those subject to the person and the institution. The Presidency therefore serves as structural power that empowers and limits the President who exercises individual power.

In the time between his defeat in the 2020 election and January 6, Trump occupied the Presidency but acted in an unofficial capacity as the defeated candidate but because he remained the President during this time his actions carried the implication of the legitimate authority of the institution of the Presidency. The weight of the office accompanies the statements and actions of the person holding it in a way increases the effect of narratives and images emanating from it. This is an effect of implicit conditioned power that is derived from "the social or cultural condition; submission to the authority of others reflects the accepted view of what the individual should do" [10, p. 29]. Implicit uses of conditioned power comprise the inconspicuous and sometimes normalized arrangements within a society that shape both the beliefs and actions of its members unknowingly to comply with a will or purpose that is not one's own. Trump used his position as President to legitimize his claims of a fraudulent election and inspire his followers to act on his behalf regardless of whether his actions constitute an official act of the Presidency or not.

Believing that the President and Presidency act in unison could explain why a group of individuals would accept demonstrably false claims as truth and expend their resources to protest and eventually attack the alleged source of the problem. In *Trump v. the United States*, the US Supreme Court ruled that "the nature of Presidential power entitles a former President to absolute immunity from criminal prosecution for actions within his conclusive and preclusive constitutional authority. And he is entitled to at least presumptive immunity from prosecution for all his official acts. There is no immunity for unofficial acts" [1]. One clear distinction made here is that the President does not always act in an official capacity while holding the office. The criteria for constituting official acts therefore becomes the focus for legal considerations, but regardless of the determination the actions of the President always carry the weight of the Presidency and this bears significant influence through implicit conditioned power as the authority of the office still holds even for unofficial acts. Trump's actions leading to January 6 while serving as President exemplify the consequences of unofficial acts taken from the institution of the Presidency. Distinguishing between the two is important for future legal cases. The implicit conditioned power of the President and Presidency will persist regardless of determinations that distinguish between official and unofficial actions of those holding the position and wielding the authority of the institution.

6.5 Power Analysis of January 6 and Its Cause

Measuring an exercise of power transpires through an examination of extensiveness, intensiveness, and comprehensiveness considered in terms of the intentionality and effectiveness of the actions. Extensiveness measures the number of subjects within the power relationship subject to a power holder or holders; there may be many power holders and many power subjects, or few, or only one of each, or any number of combinations thereof. Comprehensiveness measures either the number of domains within which the power holder commands the subject or the percentage of a power subject's total activity under control of the power holder. Domains in this sense describe the areas of interest and activity where the power relationship exists; the measure of comprehensiveness assesses either a single domain or the collective sum across all domains in which power holder and subject interact. Intensiveness describes the number of ways in which the power holder may act to win the submission of the power subject across all domains present within the relationship, and the extent to which the power holder may go in doing so while maintaining the submission of the power subject (Wrong 1979, p. 14). The intensity of the power relationship reflects its limits; the power holder can sustain the submission of the power subject in accordance with a "zone of acceptance," within which the subject complies [16].

As power holder and specific to the January 6 attack, Trump attained a low degree of extensiveness as estimates of attendance at the rally and attack number between 10,000 and 30,000 individuals [29]. In comparison, the 2017 Women's Rally saw between 3.3 and 4.6 million people protest in locations across the United States to protest Trump's election and inauguration [2]. The low number of attendees and single location relative to a comparable protest indicate low extensiveness, but the violence undertaken by the January 6 protesters as attackers demonstrate high degrees of comprehensiveness and intensiveness. Inspired by a narrative targeting core beliefs about their identity, the individuals who gathered in Washington, DC for the January 6 protest realigned their interests and exercised their individual power to fulfil a call to arms issued by President Trump. While the intentionality quotient proves high as indicated by the rhetoric and logistics involved in procuring material to build a gallows on the Capitol grounds, make shirts, secure transportation and lodging, and organize militias for action—in the case of the Proud Boys, Three Percenters, and other groups whose actions exceed this case study but warrant further review beyond the January 6 Commission Report—the effectiveness of the January 6 attack proves low as it succeeded in only delaying the certification of the electoral votes and not stopping it. A longer lens recognizes that the events of that day might have held sway in the minds of voters during the 2024 election that Trump won, but this is more attributable to that campaign than to the acute actions taken between November 3, 2020 and January 6, 2021.

6.6 Conclusion

An analysis of the events of January 6 using this power model indicates that Trump's actions on that day did not constitute an official act as President. Trump did not exercise the legitimate authority through the power of the presidency to use forces under his authority as the Commander-in-Chief to halt the Congressional certification of the electoral votes. This did not happen. Instead, Trump fabricated a false narrative based on claims of election fraud of which there remains no evidence; raised a mob of civilian supporters instead of the federal law enforcement and military services that he commanded; worked them into a frenzy by citing eschatological doom of the US, western civilization, and Christendom; and set them against the US Capitol and Congress during the official act of certifying the electoral votes in an attempt to disrupt the proceeding in an attempt to exercise power beyond the scope of the presidency and by extension to satisfy himself. On January 6, 2021, Trump failed to do so. In the days and years since, he succeeded. Trump's actions, reactions, and counteractions since leaving office on January 20, 2021 culminated in his re-election on November 5, 2024. It is safe to presume this satisfied the once again elected President Trump and constitutes an exercise of power. Analysis of the actions that Trump undertook from 2021 to 2024 exceeds the scope of this case study though it offers an opportunity for further analysis as an extension of this one. It is a significant development and warrants recognition.

Power is the ability to satisfy oneself. Holding mutually exclusive beliefs provides a mechanism for avoiding accountability as it becomes possible to justify anything to oneself; cognitive dissonance arising from antinomies of pure reason therefore provides an attractive option for individuals who idolize anything as it allows the individual to modify any belief and justify any action taken in the name of the idol [28, p. 87, 105]. This remains true no matter what the idol is. The absence of consistent beliefs makes it unlikely that an individual can ever satisfy themselves. The result is powerlessness. Belief drives identity which determines interest and the consequent exercise of power in an attempt to satisfy the original belief. Without consistency in belief, it becomes nearly impossible to satisfy oneself and therefore achieve some degree of power even if only over one's own life. The idol makes one a slave.

In his seminal work *The True Believer* [12], Hoffer recognizes the marriage between belief, identity, interest, and action and offers an explanation for why the individual grasps for such an idol despite the intransigence and futility of it all:

"Faith here is primarily a process of identification; the process by which the individual ceases to be himself and becomes part of something eternal. Faith in humanity, in posterity, in the destiny of one's religion, nation, race, party, or family—what is it but the visualization of that eternal something to which we attach the self that is about to be annihilated" (p. 64). The longing for meaning and purpose supersedes reason. Belief determines identity; identity determines interest; interest determines action that seeks to satisfy the self. Belief becomes a vulnerability when it casts a message as infallible, unfalsifiable, or inerrant, or when it elevates the messenger as a saviour or messiah. Belief as such compels submission within the power subject and breaks the identity-interest formation process by supplanting discovery and reason with certainty in dogma that paints the world in tones of a false dichotomy wherein one accepts information that reinforces the existing belief set while rejecting new information as heresy. Good information reinforces the status quo while bad information challenges it; the people who believe the former become the good guys and the latter the bad. As seen through the power theory and model, there is no such thing as the good guys. Belief is subjective and identity is contingent upon it; interest is therefore also contingent, as is the exercise of power.

References

1. Article II. S4.4.9 President Donald Trump and impeachable offenses (2024) US Congress. https://constitution.congress.gov/browse/essay/artII-S4-4-9/ALDE_00000035/. Accessed 15 Dec 2024
2. Broomfield M (2017) Women's march against Donald Trump is the largest day of protests in US history, say political scientists. Independent. January 23, 2017. https://www.independent.co.uk/news/world/americas/womens-march-anti-donald-trump-womens-rights-largest-protest-demonstration-us-history-political-scientists-a7541081.html. Accessed 10 Dec 2024
3. Burton TW (2018) The biblical story the Christian right uses to defend Trump. Vox. The biblical story the Christian right uses to defend Trump | Vox Accessed 5 Nov 2025
4. Campbell J (1968) The hero with a thousand faces. Princeton, Princeton University Press, p 53
5. Cheney K (2022) What the GOP meant when it called Jan. 6 "legitimate political discourse." Politico. https://www.politico.com/news/2022/02/15/gop-meaning-jan-6-legitimate-political-discourse-00008777. Accessed 17 Dec 2024
6. De Lamartine A (1820) L'Homme. https://www.poetica.fr/poeme-1583/alphonse-de-lamartine-homme/. Accessed 14 Feb 2025
7. Easton D (1958) The perception of authority and political change. In: Friedrich C (ed) Authority. Harvard University Press, Cambridge
8. Fischler J (2022) Pence was within 40 feet of mob inflamed by Trump, Jan. 6 committee reveals. Nebraska Examiner. https://nebraskaexaminer.com/2022/06/16/pence-was-within-40-feet-of-mob-inflamed-by-trump-jan-6-committee-reveals/. Accessed 6 Dec 2024
9. Gagne A (2024) American evangelicals for Trump: dominion, spiritual warfare, and the end times. Routledge Taylor & Francis Group, London
10. Galbraith J (1983) The anatomy of power. Houghton McMillan, Boston
11. Heinlein R (2013) Assignment in eternity. Wake Forest: Baen
12. Hoffer E (1951) The true believer. Harper Perennial, New York
13. Kruse M (2019) The escalator ride that changed America. Politico. The escalator ride that changed America—POLITICO Magazine. Accessed 18 Oct 2024

14. Lincoln A (1863) The Gettysburg address. https://www.abrahamlincolnonline.org/lincoln/speeches/gettysburg.htm. Accessed 17 Dec 2024
15. Miller P (2022) The religion of American greatness: what's wrong with Christian nationalism. Intervarsity Press, Downers Grove
16. Partridge P (1963) Some notes on the concept of power. Political studies. 11(2): 107–125. https://doi.org/10.1111/j.1467-9248.1963.tb01054.x.. Accessed 23 Sep 2020
17. Pilat D, Sekoul D (2021) Illusory truth effect. The decision lab. Retrieved Oct 23, 2024, from https://thedecisionlab.com/biases/illusory-truth-effect. Illusory truth effect—the decision lab. Accessed 24 Oct 2024
18. Seidel A (2021) The founding myth: why Christian nationalism is Un-American. Union Square & Co, New York
19. Seidel A (2022) Christian nationalism and the capitol insurrection. Written testimony of Andrew L. Seidel of the freedom from religion foundation on the role Christian nationalism played in the lead up to and during the attack of January 6. https://www.govinfo.gov/content/pkg/GPO-J6-DOC-CTRL0000062431/pdf/GPO-J6-DOC-CTRL0000062431.pdf. Accessed 22 Nov 2024
20. Separation of Powers and Checks and Balances | Constitution Annotated | Congress.gov | Library of Congress
21. Simon M, Sidner S (2021) Decoding the extremist symbols and groups at the Capitol Hill insurrection. CNN. January 11, 2021. https://www.cnn.com/2021/01/09/us/capitol-hill-insurrection-extremist-flags-soh/index.html. Accessed 6 Dec 2024
22. Stacy J, Whitehead A (2024) How we can confront the myths of January 6 and intensifying Christian nationalism. Time. https://time.com/6552370/january-6-myths-christian-nationalism/. Accessed 17 Dec 2024
23. Stewart K (2019) The power worshippers: inside the dangerous rise of religious nationalism. Bloomsbury Publishing, New York
24. Stewart K (2022) Christian nationalism is one of Trump's most powerful weapons. The New York Times. https://www.nytimes.com/2022/01/06/opinion/jan-6-christian-nationalism.html. Accessed 17 Dec 2024
25. Trump D (2021) Transcript of Trump's speech at rally before US Capitol riot. Associated Press. Transcript of Trump's speech at rally before US Capitol riot | AP News. Accessed 18 Oct 2024
26. Wheeler R (2021) Trump's judicial campaign to upend the 2020 election: a failure, but not a wipe-out. Brookings Institute. https://www.brookings.edu/articles/trumps-judicial-campaign-to-upend-the-2020-election-a-failure-but-not-a-wipe-out/. Accessed 4 Dec 2025
27. Wrong D (1979) Power: Its Forms, Bases, and Uses. Transaction Publishers, London and New Brunswick
28. Zizek S (2008) Violence. Picador, New York
29. Zou and Logan (2022) https://www.latimes.com/politics/story/2022-01-05/by-the-numbers-jan-6-anniversary. Accessed 10 Dec 2024

Chapter 7
Conclusion

> The way is for ever nameless
> Though the uncarved block is small
> No one in the world dare claim its allegiance
> Should lords and princes be able to hold fast to it
> The myriad creatures will submit of their own accord,
> Heaven and earth will unite and sweet dew will fall,
> And the people will be equitable, though no one so decrees.
> Only when it is cut are there names.
> As soon as there are names
> One ought to know that it is time to stop.
> Knowing when to stop one can be free from danger.
> The way is to the world as the River and the Sea are to rivulets and streams.
>
> Lao Tzu, XXXII [33]

> All the rivers run into the sea; yet the sea is not full; unto the place from whence the rivers come, thither they return again.
>
> Ecclesiastes 1:7 [30]

The research that became this book began in February 2017. The world that existed then, at the time when the first keystrokes began giving new life to old ideas, no longer exists. Operation Odyssey Lightning in Libya had concluded only two months prior. Russia held Crimea in defiance against the sanctions levied against it. The United States remained a signatory to the Paris Agreement and inaugurated Donald Trump as its President a few weeks before. In 2022, the Libyan Civil War continues—Libya remains a failed state embroiled in civil war that shows little prospect of abating [12]. Russia still holds Crimea and invaded Ukraine in February 2022 invoking the same justification "that Ukrainians are 'one people' with Russians, living in a failing state controlled by Western forces determined to divide and conquer the post-Soviet

K. L. Wedekind, *Constructing Power: A Critical and Constructivist Theory Using the American Case*, Advanced Sciences and Technologies for Security Applications,
https://doi.org/10.1007/978-3-031-84893-3_7

world" remains a key justification for what might follow, even if the practical goal for Russia reflects a desire to assert itself against the perception or reality of Western encroachment by denying the North Atlantic Treaty Organization (NATO) additional and significant frontage on its western border [25, 32]. During the first year of the Trump administration, the United States declared its intent to withdraw from the Paris Agreement and actually did so during the final year of that administration on November 4, 2020—the day after the 2020 United States Presidential Election, in which Joseph Biden defeated Donald Trump [24]. President Biden announced on his first day in office that the United States would re-join the Agreement, which it did on February 19, 2021 [7].

In 2021, the transition between the Trump and Biden administrations did not transpire peacefully for the second time in the United States, with the first having led to the American Civil War, as insurrectionists stormed the United States Capitol on January 6, 2021 to disrupt the certification of the results from the 2020 Presidential election [2]. The America First ideology espoused by Trump and its progeny of conspiracy theories gave credence to the belief among many Republicans that the 2020 election contained widespread voter fraud resulting in an incorrect count in the electoral college—this claim is false [6, 9]. It did prove highly effective in motivating supporters to attempt an insurrection—all at the urging of the outgoing President and his fellow conspirators within Congress, the media, and elsewhere—on January 6, 2021 in the United States Capitol as Congress set itself to the task of certifying the results of the election [4]. The January 6 Commission concluded its investigation and released its report outlining how the incident developed, unfolded, and who bears responsibility for the deliberate attempt to destroy the democratic process in the United States [3].

In November 2024, the citizens of the United States re-elected Trump to the Presidency. The oft stated adage that "elections have consequences" always holds true and this administration change will remain consistent with this sentiment. As of this writing in December 2024, it suffices to recognize that the re-election of the principle antagonist to the US Presidency complicates the legacy of January 6, those who participated, and the institutions that were attacked. It is also likely that American global leadership will tilt away from many policies established during the Biden presidency, which will continue the trend of waffling on various issues like climate change every four to eight years.

In December 2019, a novel coronavirus emerged and has caused an ongoing global pandemic in which 5.3 million individuals have died because of the illness [21, 39]. Within one year, scientists collaborated to produce several vaccines that have proven effective in reducing the severity of illness and likelihood of death [11]. Despite its efficacy and safety, millions remain unvaccinated with many citing conspiracies as evidence [5]. An examination of vaccine reluctance in terms of power could provide insight into individual and group behaviour; the effects of conditioned power as the operative instrument in shaping actions and beliefs entices the researcher into future works, especially focused upon the interplay between power and trust.

In August 2021, the United States withdrew from Afghanistan and ended its war there. Over its course, the United States war in Afghanistan caused 243,000 deaths and cost $2.3 trillion [35]. The entire endeavour meandered from its origin through a desert back to whence it came, leaving none the same as once they were. The United States war in Afghanistan exemplifies the undercurrents of complexity within power; reducing it from monolith into an approachable form will require much consideration through the lens presented in this work.

On February 24, 2022, Russia invaded Ukraine under the pretences of a "special military operation" [1]. Lethal military aid from NATO and its member states combined with economic sanctions by the European Union, United States, United Kingdom, other Western states, and Japan have bolstered Ukrainian defences and will likely cripple the Russian economy for years, if not decades [31]. Ukraine continues to defend itself across its territory against Russian forces. On February 27, Putin placed Russian nuclear forces into a "special mode of combat duty," the meaning of which remains unclear but signifies an escalation in response to Ukrainian defence and its support from the West along with the sanctions by the West against Russia [29]. Future research into the causes and consequences of the 2022 Russia—Ukraine War will do well to view the conflict through the long lens of history and the power model presented in this book, especially the case study in Chap. 3, which presents the 2014 conflict and annexation as the second act of this real human tragedy playing upon the world's stage, with the prelude and first act having transpired over a thousand years and the thirty since the dissolution of the Soviet Union, respectively.

Within each change since 2017 and in all that transpired before and will hereafter, power serves as the primary driver. Power wrought the change that has transformed the world of 2017 into present day. While much appears as it was, little remains of the identities that defined who each person believed themselves to be—these passing manifestations of identities become memories of who one was at a certain place in a given time, lost to the relentless procession of power through time and space. Power proceeds to shape by breaking, binding, and breaking again. The principles of power remain consistent across time and space while the manifestations and consequences reflect the specifics particular to the social interactions that unremittently continue—"So we beat on, boats against the current, borne back ceaselessly into the past" [17].

The questions driving this work ask both theoretical and practical questions about power. The theoretical questions seek to understand what power is and how it functions; from where power comes and how one exercises it; and how identity, interest, and power relate to each other. Chapter 1 of this book answers these theoretical questions by outlining the nature of power and its elements; identifying the dependency between identity, interest, and the exercise of power; and offering a new power model that can explain the process through which every intersubjective social-power relation transpires. The practical questions about power inquire into how United States national identity, interest, and exercises of power have transpired, which Chap. 2 through Chap. 6 reviewed. This chapter summarizes the findings and draws conclusions about power in theory and praxis.

7.1 A New Definition of Power

Power is the ability to satisfy oneself. This definition of power unites identity, interest, and the exercise of power into a single concept. The word choice is key. Ability here refers to the sources, instruments, and forms of power in association with one another. The purpose of this association is to possess the means of achieving conditions that satisfy, which requires the existence of a set of conditions that please—an interest. Interest does not exist independently, but rather depends upon identity for its characterization. Identity forms from interaction with the other and distinguishes these two agents from one another, including who each one is, what each one wants, and what either one can do about it. The use of oneself in this definition speaks of both the individual agent and the collective, which implies that the latter must possess identity and interest beyond the individuals comprising it.

This definition of power is new and distils the essence of this work into a single but complicated sentence. It captures the essence of this research, though many implications within it require explanation for such a brief definition to reflect completely the theory outlined within these pages. The three critical elements of this definition reflect the gravities that bind this power theory—identity, interest, and power. The use of the self as the direct object implies the existence of an identity and focuses the attention on it. Identity can only exist because of interaction with the other, thereby implying the interaction that initiates role-taking and alter-casting that defines self and other; that is, ego and alter, I and thou [22, p. 6]. Each agent party to the engagement brings their own conceptions of self, the endogenous perception, only to have the other overrule it with an exogenous perspective that discounts perceptions of self in favour of an identity constructed from the bits of information that present within the interaction; each agent is neither who they think themselves to be nor who the other thinks them to be, but rather each is who they think the other thinks they are [37, p. 404]. Every such engagement creates a unique identity for both agents with each distinct interaction; the relentless passage of time and subsequent engagements with other external agents changes both the original agents, even if only a little but enough to demand a reconfiguration of identity relative to one another [20, p. 39]. Identity therefore emanates from self as a result of the interaction with the other. The formation of identity produces an interest, defined as the set of criteria or circumstances particular to the relation with the other that upon achievement would satisfy the self. The creation of identity through interaction causes the formation of an interest; identity and interest both derive their character from the way in which the self understands itself relative to the other in combination with its understanding of the other [38, pp. 224, 328]. This dynamic between identity and interest produces subjective understanding that can account for one reason as to why an agent may not act in its own best interest. The requirement for satisfaction in this definition of power implies the existence of interest, which requires an identity formed through interaction with the other. The ability by which an agent can satisfy itself arises from the exercise of power against the other by either overcoming or resisting the reciprocal exercise undertaken by the other against the agent. Power emanates from the sources of personality, property, and organization; takes shape as an instrument reflecting either condign, conditioned, or compensatory power; and then manifests in the form of either force, authority, persuasion, or manipulation [19, p. 55, 40, pp. 24, 28, 32, 35]. The power construction process may incorporate multiple sources, instruments, and forms into a single exercise of power, which must display intentionality and prove effective to constitute an exercise of power; effectiveness reflects extensiveness, comprehensiveness, and intensiveness, which correlate to the number of agents and the number of activities influenced, and the degree to which the power holder may push the subject while sustaining the relation [13, p. 160, 27, p. 118, 40, p. 14]. The possession of power can constitute an exercise of power if the agent has accumulated it for the purpose of deterring particular actions by the other and succeeds in doing so without exercising the power possessed against the other; when so cultivated, latent power meets the criteria of intentionality and effectiveness to constitute an exercise of power. Power is the ability to satisfy oneself; this definition implies the existence of self and other, their respective identities and interest, and abilities to exercise power against one another either to overcome or to resist.

The satisfaction of oneself recognizes the individuals inherent within every collective. Each individual possesses an identity and interest relative to the other, which includes a unique identity and corresponding interest that manifests relative to every collective of which the individual counts themselves a part. A submission of individual identity and interest precedes the formation of the collective and sustains it; the collective formulates its own identity, interest, and power through the consolidation of its constituents into an amalgamation reflecting its unique self relative to its parts. Existing within a collective does not dissolve the individual but rather absorbs them; the individual must submit to the collective so that it may exist, but the individual retains their own identity and interest while existing both as a part of the collective and an individual independent from it. This reconciliation of individual and collective permits for the reduction of social relations to transpire between two and only two agents while allowing for multiple interactions to occur concurrently [23, 26, p. 231, 34, p. 167].

Every social relation is a power relation. The term intersubjective social-power relation captures the dynamic as a comprehensive concept that recognizes relations as transpiring between two and only two subjects, the sameness of social and power as not different types of relations but as identical types, and the relation itself as the catalyst of the entire process that unfolds through engagement between agents. Discussion outlining the reduction of relations to two and only two agents appears in the previous paragraphs. The equivalency between social and power relations recognizes that the state of nature for all is conflict with the other. This conflict begins upon birth and continues throughout life in myriad forms until death, persisting even within the social contracts that exist within each respective social collective. Power brings individuals together and binds them to one another; this coalescing forms the ties that bind through the perception of common or shared beliefs about self, other, and the world in which they exist. In this light, power presents itself as the duality represented by the Tao within the power model—power creates and destroys, binds and breaks, gives life and takes it, as a value neutral gravity or energy that manifests in every social relation, which draws its moral and ethical value from the beliefs and actions of the agents party to the engagement [8, pp. 220–30, 15, p. 28, 16, p. 292, 18, p. 94].

Each interaction constitutes a social relation between subjects wherein power serves as the operative and determinative gravity [10, p. 195, 38, p. 1]. Power drives each relation along the same pathway, but each engagement transpires differently by reflecting the unique understanding of self and other, what each desires, and the actions that each takes to satisfy themselves—identity, interest, and the exercise of power, respectively. The process outlined within the power model follows the same steps within the relation; interaction between agents creates their identities relative to the other along with a corresponding interest that constitutes belief about what set of circumstances would satisfy each that leads to the actions taken in pursuit of the self-determined goals believed to satisfy upon attainment. The power model can serve as heuristic framework designed to account for the multitude and diversity of factors influencing the way in which each engagement transpires and its consequent outcome, making the model both descriptive of the process and prescriptive of the outcome when able to consider the specific parameters of a given engagement. The following sections answer critical questions about the nature of power—what it is, how it works, and why.

7.2 The Relationship Between Identity, Interest, and the Exercise of Power

Within this new power model, the interaction creates the identity and interest of the two agents participating in the engagement and facilitates the exercise of power between them. Identity takes its shape through role-taking and alter-casting that defines who each believes themselves and the other to be; interest arises from the identity formed within the relation, and the exercise of power provides the means and ways through which each seeks to satisfy themselves. Implicit in the definition of power as the ability to satisfy oneself is the existence of the self and its identity that can exist only relative to the other. The existence of identity corresponds to an interest, which further corresponds to the exercise of power. These interrelated elements comprise the gravity that binds and breaks all relations attributable to the concept of power. This power model, as supported by the theory in the Chap. 1 and subsequent case studies, constitute an original contribution to knowledge. The model provides a lens through which future research can view and analyse both new and historical power relations between agents, whether state or non-state actors. Applying the theoretical findings of this work and its model to other case studies gives it a life beyond the dissertation (Fig. 7.1).

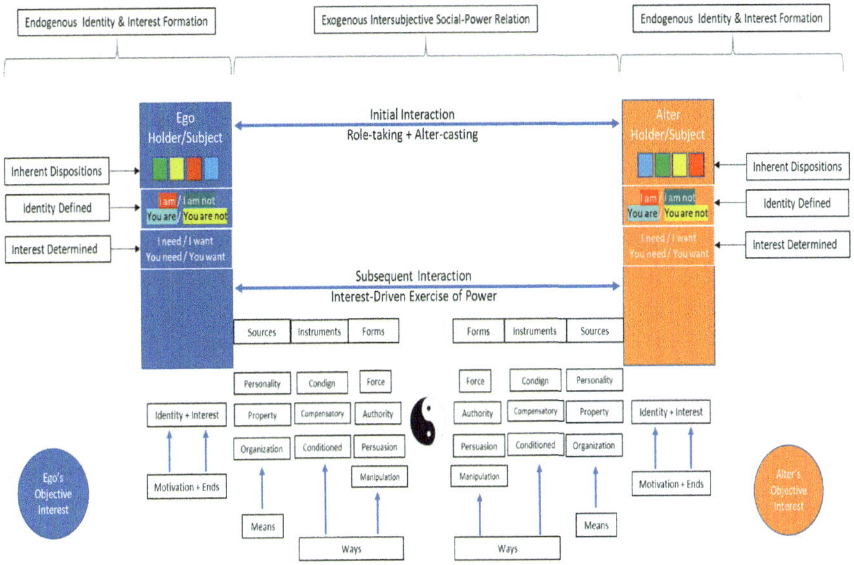

Fig. 7.1 New power model (Author's own)

7.3 The Association of Sources, Instruments, and Forms of Power

Within the intersubjective social-power relation, agents utilize power to seek satisfaction of their respective interest derived from their identity relative to the other. The identity and interest formation process comprise the initial stages of power and drive the acquisition and exercise of power, which emanates from sources, takes shape through an instrument, and manifests in a form. The relationship between the sources, instruments, and forms reflects the strategic, operational, and tactical paradigm, respectively. Sources define what is possible, while the instruments provide a mechanism for converting the possible into action that appears as a form. The entire process constitutes the exercise of power.

7.4 The Sources of Power

Personality, property, and organization comprise the sources of power. Personality "is the quality of physique, mind, speech, moral certainty, or other personal trait that gives access to one or more of the instruments of power" [19, p. 6]. Personality includes all aspects of personhood; the physical and psychological, including how one appears, what one believes, and how one conducts themselves. The effects of the law, education, and history influence personality, but these pressures constitute intellectual property and therefore arise as sources of power from property, which refers to materiel, capital, and knowledge—the possession of resources, whether physical, financial, intellectual, or otherwise, that allows the owner to bestow boons or tantalize the other by withholding the same [19, pp. 6–7]. Organization is "a number of persons or groups...united for some purpose or work" [19, p. 55]. Organization is in this sense is an action and an entity; the process of organizing can operate as a source of power, just as can the machinations of an established organization. Organization provides a way for associating personality and property with interest and identity that allows for subsequent exercises of power to achieve the intentionality and effectiveness quotients.

For the state, personality appears as a concept of national identity while property accounts for all that it owns, possesses, and controls in terms of materiel, capital, and knowledge. Organization presents as government and the processes that constitute governing; that is, the thing become and becoming that gives the state the ability to arrange its property and present its personality in its relations with others, though the initial interaction with another state determines the character of its national identity relative to the other. The United States possesses its own self-perceived identity that exists independently from the other, but upon interaction with the other this identity fades away through role-taking and alter-casting that transpires between the United States and a given other. Each derives a unique version of itself from the interaction that determines the interest of each within the engagement. National identity understood as personality as a source of power reflects manifestation of identity created through the engagement and not the independent self-perception. The property available to the United States includes its equipment, supplies, capital, technology, research and development, and other assets including population. While the United States does not own its people in the legal sense, it does require that its people do serve even if under the guise of voluntary servitude, which constitutes the submission of self to other; that is, people submit to the law that the state creates and despite the ways in which the state creates it, the submission of the people to the state transpires as a necessary part of the social contract. The people consent to government through submission to the law. Organization describes the ways in which the United States arrays its assets and sets them to its purpose understood as national interest. The machines of government organize state property in ways that make it usable in pursuit of the goals that collectively define national interest.

7.5 The Instruments of Power

The instruments of power are the ways in which an entity wields or enforces power [19, p. 4]. Instruments transform the energy cultivated from the sources of power into a design that will manifest into a singular form. Through this process chain, the exercise of power transpires. Three instruments exist—condign, compensatory, and conditioned—and account for the many ways in which sources convert to forms.

Condign power "wins submission by the ability to impose an alternative to the preferences of the individual or group that is sufficiently unpleasant or painful so that these preferences are abandoned" [19, p. 4]. The act or intent to punish either physically or psychologically characterizes condign power. Compensatory power "wins submission by the offer of affirmative reward—by giving of something of value to the individual so submitting" [19, p. 5]. The act of giving along with demonstrating the intent to give constitute compensatory power, just as do the withholding of rewards or displaying an intent to withhold rewards. Conditioned power "is exercised by changing belief" [19, p. 5]. Conditioned power focuses on changing the perceptions and beliefs of the agent about either the environment in which they exist or the agent itself. Each of the three instruments seek to change the behaviour of the other by attacking either the agent or its beliefs; an agent submits based on their belief that to do so is the best course of action available to them when reconciling their perception of the situation in which they find themselves and their own interest.

The state wields the instruments of power through the actions that it takes to implement policy. The instruments characterize the type of activity that the state will use to satisfy its interest and identity comprising the self. Instruments facilitate the conversion of sources into forms. For the United States, the instrument of condign power bears a close relationship with its military and the use of force while compensatory power reflects economic policy. Conditioned power relies on the use of information and diplomacy. The instruments of power reflect the case studies; while the sources define what is possible and the forms present power manifest, the instruments of power reveal the character of power and by extension insight into the interest it serves in a way that reveals the identity of the agent exercising power.

7.6 The Forms of Power

The forms of power constitute the actual application of power between two agents. Four forms of power exist—force, manipulation, persuasion, and authority. Each form of power may emanate from any of the three instruments. Force "refers most commonly to physical or biological force: the creation of physical obstacles restricting the freedom of another, the infliction of bodily pain or injury including the destruction of life itself, and the frustration of basic biological needs which must be satisfied if the capacity for voluntary choice and action is to remain unimpaired" [40, p. 25]. Force exists in two subtypes, violent and non-violent, which may target either the physical body or psyche of the power subject through either the application or withholding of force. Force and condign power maintain a close association, though both compensatory and conditioned power manifest through force as a form of power. Compensatory power becomes force when a dependency exists between agents such that the giving or withholding of rewards controls the behaviour of one agent especially in circumstances that compels acting against interest and identity as a means of sustaining the relationship; conditioned power that convinces an agent that acting against its own interest is actually in its interest constitutes force. The removal of choice or appearance thereof plays a determinative role in exercises constituting the use of force.

Persuasion occurs "when the power holder presents arguments, appeals, or exhortations to the power subject, and the power subject, after independently evaluating their content in light of his own values and goals, accepts the power holder's communication as the basis of his own behaviour, the power holder has successfully persuaded [the power subject]" [40, p. 32]. Manipulation is the form of power that occurs "when the power holder conceals his intent from the power subject—that is, the intended effect he wishes to produce, he is attempting to manipulate the latter;" or, stated another way, "any deliberate and successful effort to influence the response of another where the desired response has not been explicitly communicated to the other constitutes manipulation" [40, p. 28]. The critical factor of manipulation is the use of deceit, which distinguishes manipulation from persuasion [36, p. 130]. Persuasion involves using arguments to convince the other to modify their behaviour, while manipulation denies such a choice to the other and appeals to notions that result in the behaviour change desired by the manipulator.

The fourth form of power is authority, which "is the successful ordering or forbidding…any and all command-obedience relations…are examples of authority" [40, p. 35]. There are five subtypes that distinguish between the kinds of authority relations—coercive, induced, legitimate, competent, and personal authority. In each subtype, the power holder wields the subtype of authority exercises it to gain the submission of the power subject to his will. Coercive authority occurs when the power holder, having convinced the power subject of both his capability and willingness to use force to induce submission, obtains the compliance of the power subject [40, p. 41]. Induced authority transpires when the power holder offers or provides rewards to the power subject to gain his submission [40, p. 44]. Legitimate authority exists when "the power holder possesses an acknowledged right to command and the power subject an acknowledged obligation to obey" [40, p. 49]. Competent authority results from the power subject's recognition of the power holder's possession of specialized knowledge or skill, and "obeys the directives of the authority [power holder] out of belief in the authority's superior competence or expertise to decide which actions will best serve the [power] subject's interests and goals" [40, p. 53]. Personal authority transpires when "the subject obeys out of a desire to please or serve another person the power holder solely because of the latter's personal qualities" [40, p. 60].

7.7 The National Identity, Interest, and Power of the United States

The four case studies presented in this book indicate that a causal relationship exists between identity, interest, and the exercise of power. With each, identity and interest comprise the causal element while the exercise of power serves as the effect. The identity formation process begins anew with each interaction between the parties to a given relation, which causes the interest of each to change at a rate that correlates to the frequency of the interactions within the relation. To actualize the interest and satisfy the identity, an agent exercises power against the other while the other resists through reciprocal or countervailing exercises of power. The result proves binary. One agent overcomes the other, which makes the former the power holder and the latter the power subject. Engagements that transpire over an extended period will remain suspended until one party to the relation achieves its interest. Engagements in which the interests of two agents align follow this same process; in such cases, the disposition of the relation may prove amicable. The source of the power relation therefore lies with identity and how the agents party to the relation understand themselves relative to one another. This dynamic compels the emphasis in this research upon methodological individualism and collectivism. The individualist perspective can offer insight into the behaviour of one person, which may prove appropriate for a particular analysis but an individual never exists beyond the strictures of collective organizations. The Hobbesian state of nature never shapes a person as each is borne into a world that precedes them and begins to influence their behaviour from birth, never ceasing. Analysis that considers only the individual neglects the influence of the collectives on their behaviour, while focusing only upon the influence of the collective neglects the experiences that cultivate the ingrained identity of the individual. Within engagements between collectives, power relations exist within each between the individuals comprising the organization and between the individuals and the inertia of the organization itself. As each relation is a power relation, those that transpire between members of a single group follow the same process as that which transpires between their organization and another; similarly, the individual within an organization continuously engages in a power relation with the organization itself, or at least the individual's understanding of it. Both methodological perspectives warrant consideration when seeking to understand power due to its ubiquity and the multitude of manifestations within a given episode; given the magnitude of power relations, analysis must select certain aspects upon which it will focus rather than attempt to encapsulate the entirety of an episode by including all the collectives and individuals contributing to the fray.

This book examined a collective agent, the United States, and the power formation process that shaped its engagement with other collectives through three case studies following an examination of the role of American exceptionalism as a core belief and therefore driver of the United States' national identity, interest, and exercises of power. In the end, it becomes clear that national identity is as contingent upon belief as individual identity. National identity, and the national identity of the United States, proves transient but remains relative to the belief set of the preponderance of individuals holding authority—where legitimate maintains primacy but also through personal, induced, coercive, and competent authority. From this mess, a visage clearer through peripheral vision rather than direct observation arises that serves as a national interest but reflects the national identity of those holding authority rather than the entirety of the collective nation.

Chapter 2 examined American national identity and the apparent cognitive dissonance caused by the conflict between two competing ideologies—American Exceptionalism and America First—and its effect upon the national identity and interest of the United States and its willingness to exercise power. This case study examined the identity and interest formation process that precedes the exercise of power and revealed a crisis of conscience about the collective identity of the United States held in the minds of American citizens. Identity serves as the catalyst for the development of interest, which demands the exercise of power in its service. A fractured identity creates confusion over what conditions define the interest of the agent, which produces exercises of power that bear no relation to the interest and identity of the agent. The identity formation process itself then becomes the grounds for the power formation process to occur. The United States has derived its national identity from the competing principles embraced by American Exceptionalism and America First since the arrival of Europeans in North America. The idea that the United States possesses a special mission relative to the other states in the international community has driven its exercise of power for centuries, as evidenced broadly by Manifest Destiny, the Monroe Doctrine and Roosevelt Corollary, Wilson's Fourteen Points, the foundation of the United Nations, the influence of the domino theory during the Cold War, the Gulf War, and the Global War on Terrorism. The first three case studies offer specific examples in which the United States completed the power construction process and exercised power; this fourth case study focused on why the United States would do so and finds the answer within its identity. The United States exercised its power because the leaders holding positions of legitimate authority within its government at the time determined that the national interest of the United States necessitated certain actions to satisfy the role-taking and alter-casting that formed its collective identity within each episode.

In Chap. 3, the first case study examined the conflict between the United States and ISIS-L in the vicinity of Sirte, Libya in 2016. The United States exercised condign power through the use of force, including both physical and psychological violence, that finds its source in organization and property. This exercise of condign power proved both intentional and effective; public statements from official spokespersons provided evidence of intent, while the elimination of ISIS-L from Sirte using airstrikes in support of ground operations undertaken by the Government of National Accord achieved effectiveness with high degrees of intensiveness and comprehensiveness but demonstrated lesser extensiveness due to the limited scope of the operation. The resulting destruction of ISIS-L in Sirte proved a tactical success but failed to produce any subsequent strategic victory in the form of stabilization within Libya, a failed state, or across the North African region. The United States reluctance to reconstitute Libya despite its role in setting determinative conditions for sustained civil war establishes a new precedent wherein the United States views its interest in terms of itself alone as opposed to aligning its national interest with international interest and its respective contribution to global stability; that is, Operation Odyssey Lightning reaffirms the distinction between the respective identities and interests of states as individuals comprising the collective international community. This case study shows that the United States defines its global responsibilities relative to its own national interest rather than a common interest shared across the collective international body politic.

The second case study in Chap. 4 reviewed the annexation of Crimea by the Russian Federation in 2014 and the economic sanctions that the United States levied against Russia in response. This case study considered the role of Russian identity and national interest as the motivation and catalyst for its exercise of power as well as the sources, instruments, and forms of power that Russia integrated into its decades-long strategy; juxtaposed with the United States and its lack of identity ties and consequent waning interest in the affair issued economic sanctions in response to the Russian annexation. The comprehensiveness, intensiveness, and extensiveness of the Russian exercise of power proved great while the United States attempted exercise of power proved little. The Russian exercise of power combined multiple sources, instruments, and forms into an operation that spanned decades demonstrating a patient and enduring intentionality. Effectiveness reflecting high degrees of comprehensiveness, intensiveness, and extensiveness revealed itself through Russia's successful annexation and hold onto Crimea against economic sanctions and public condemnation from the United States and others in the international community. The corresponding reaction by the United States to the Russian action lacked any discernible measure of effectiveness produced by economic sanctions and rhetoric. In the action-reaction-counteraction cycle, Russia dominated the initial action and withstood the reaction from the United States and others, which demonstrated both the power to do and the power to resist. The actions by the United States may have achieved their intended effect to meet the intentionality and effectiveness quotient constituting its reaction as an exercise of power, but its extensiveness proved very little given the number of individuals subject to economic sanction with the comprehensiveness and intensiveness registering almost not at all. Russia defeated Ukraine to annex Crimea, but also withstood the United States and the international community opposing its exercise of power.

In Chap. 5, the third case study considered the effects of the United States' abrupt abandonment in 2016 of the Paris Agreement of 2015 following decades of advocacy for such an international climate change mitigation agreement. This case study showed the intentionality and effectiveness of the United States' diplomatic efforts from 1992 to 2015 as indicated by the achievement of the Paris Agreement, though its abandonment in 2017 demonstrated low degrees of comprehensiveness, intensiveness, and extensiveness within the American electorate. The United States had exercised persuasive conditioned power for decades during negotiations towards an international climate change agreement and dedicated significant effort, including direction negotiations by the President of the United States with multiple heads of state or government, in the final effort to reach an agreement. The negotiations constitute an exercise of power for meeting the intentionality and effectiveness quotients while demonstrating extensiveness, comprehensiveness, and intensiveness as 192 of 197 states ratified the agreement and set out to meet its parameters. The 2016 general election in the United States would usher in a new administration that declared its intent to withdraw from the Paris Agreement, lowering the total number to 191. The simple act of abandoning an agreement that the United States had championed during the final stages of its development devalued United States leadership internationally and its ability to exercise persuasive conditioned power during subsequent international negotiations. Inconsistency between political administrations in the United States casts it as an unreliable individual member of the community of states. This inconsistency emanates from polarization in the domestic politics of the United States and reflects the cognitive dissonance present within American collective identity for the past twenty years.

The fourth case study in Chap. 6 examined how power functioned during the leadup to and during the January 6 attack on the US Capitol. President Trump leveraged his ability to influence belief, identity, interest, and the exercise of power by his supporters to inspire their actions that disrupted and delayed but did not stop the certification of the electoral college that sealed his defeat in the 2020 election. Trump operated across multiple, mutually reinforcing lines of effort that focused on changing what people believed about the results of the election and capitalizing on many of his supporters' predisposition for distrusting the institutions of government despite the absence of evidence. Role-taking and alter-casting figured prominently as Trump embraced the self-fulfilling prophecy that he had manufactured over years as the persecuted saviour sacrificing himself for his ideals and his disciples. The causal relation between belief shaping identity, determining interest, and drive the exercise of power becomes clearest in this case study and supports the power theory, model, and definition presented within this work.

7.8 The Importance and Significance of This Research

The ubiquity of power and its determinative effect on the events that shape the world, from the least of interactions to the greatest, makes this research important. Power shapes relationships and determines their disposition, which necessarily comprises two and only two agents that may be either individuals or collectives; within the relation, the exercise of power determines which agent becomes the power holder and the other the power subject. The theory here offered outlines the power construction process that transpires as agents initiate a relationship and engage with one another both within single episodes and within the greater context of the relationship that exists between them. The identity formation process begins anew with each interaction. A corresponding interest forms based on the circumstances characterizing the time and space of the environment. Power then serves as the operative force at play within the relation. Power emanates from sources, takes shape as an instrument, and then manifests in a form; the power construction process may incorporate multiple sources, instruments, and forms into a single exercise of power. An exercise must prove intentional and effective to constitute an exercise of power; effectiveness reflects extensiveness, comprehensiveness, and intensiveness, which correlate to the number of agents and the number of activities influenced, and the degree to which the power holder may push the subject while sustaining the relation.

The intersubjective social power relation, which describes the engagement itself as opposed to the entire power construction process, exists between every two agents as each interacts with the other. Engagements that transpire between two individuals or two collectives reflects this dictum purely while those that occur between an individual and a collective require reducing the collective to the individuals representing it for proper analysis. This perspective lends itself to moderate methodological holism and its position that both individualist and collectivist positions may prove valid depending upon the circumstances of the situation under examination [41].

Power is both complex and complicated. Complexity arises from the several phases and steps within each that define the power construction process and the iterative nature of the process that depends upon feedback that arises from re-engagement with previously encountered agents; complications emanate from the perspectives of the individuals engaging with one another through the process and the potential for the misalignment of identity and interest with the exercise of power. If every engagement between two agents constitutes a power relation then the actions within the relation that demonstrate effectiveness and intentionality indicates that such power relations further constitute complex adaptive systems, which necessarily "contain adaptive components and capacities…[that] allow systems to change and evolve over time in response to feedbacks and changes in the system context. This means that complex adaptive systems have a memory and the capacity to learn from previous responses and configurations and thus, influence and shape current and future system trajectories" [14, p. 312, 28, p. 2]. Power relations understood as complex adaptive systems "contain no a priori assumptions about key variables, emphasize nonlinear causal effects between and within systems, and view system equilibrium as multiple, temporary, and moving" [14, p. 312, 28, p. 2]. The power relations between collective agents that reflect this description closest exists between states, transpires within the international order, and occurs along multiple, nonlinear lines of effort that the power theory seeks to illuminate. The condition wherein states do not declare a state of war but engage in warfare across all domains of national power proves most interesting; while the sources, instruments, and forms of power prove finite in theory and limited in reality, the ways in which the state as agent may combine them reflects the uniqueness of the agents, their engagement, and the circumstances characterizing their interaction. This work offers a heuristic framework that can account for the multitude and diversity of factors influencing the way in which each engagement transpires and its consequent outcome. The case studies demonstrate the power construction process and suggest that what appears as an exercise of power might not meet the criteria for actually being one, especially when compared to exercises of power that demonstrate intentionality and effectiveness.

7.9 The Perpetual Search for Meaning and Understanding

This book offers a way to understand how power functions and why agents choose to exercise their power by accounting for the causal effects of identity and interest coupled with an accounting of the way in which power functions. Coalescing these aspects of power theory into a singular model presents a comprehensive way to understand power and the role it plays in shaping the world through the intersubjective relations that characterize it. The fundamental principles upon which this research stands and that constitutes an original contribution to knowledge recognize that every relation is a power relation that transpires between two and only agents, which may be either individuals or collectives, who operate within an environment that structures their respective beliefs about self and other within the context of their interaction. Beliefs held by the agents about self and other determine identity uniquely within each interaction transpiring in space and time, even when between the same two agents. Identity produces a set of ideas and goals that comprise an interest that requires subsequent action through the exercise of power, which emanates from sources, finds its disposition in an instrument, and manifests as a form. The exercise of power corresponds to the subjective interest of the agent and must reflect intentionality while achieving effectiveness in terms of comprehensiveness, intensiveness, and extensiveness. The exercise of power may project upon another or resist such projection as an aspiring unstoppable force or immovable object keen to remain as such, respectively. These principles inspire the power model that recognizes power as the key force that shapes each agent and every social relation while also showing how power functions between two agents within their every interaction. In this light, power constructs the world and everything in it ceaselessly; to begin to understand the world, one must first understand power.

References

1. Al Jazeera (2022) Russian forces launch full-scale invasion of Ukraine. February 24, 2022. https://www.aljazeera.com/news/2022/2/24/putin-orders-military-operations-in-eastern-ukraine-as-un-meets
2. American Civil War Museum (2016) The election of 1860 and the transfer of power. October 27, 2016. https://acwm.org/blog/election-1860-and-transfer-power/. Accessed 15 Dec 2021
3. Amiri F, Jalonick MC (2021) House votes to hold Mark Meadows in contempt in Jan. 6 probe. December 15, 2021. https://apnews.com/article/steve-bannon-capitol-siege-subpoenas-mark-meadows-congress-bd6ea8ca859228c89e875e7a7c47317f. Accessed 15 Dec 2021
4. Associated Press (2021) Capitol Siege. https://apnews.com/hub/capitol-siege. Accessed 15 Dec 2021
5. Bertin P, Nera K, Delouvée S (2020) Conspiracy beliefs, rejection of vaccination, and support for hydroxychloroquine: a conceptual replication-extension in the COVID-19 pandemic context. Front Psychol 11:565128. https://doi.org/10.3389/fpsyg.2020.565128. https://www.frontiersin.org/articles/10.3389/fpsyg.2020.565128/full. Accessed 15 Dec 2021

6. Blake A (2021) The most brutal debunking of Trump's fraud claims yet—from Republicans. The Washington Post; June 24, 2021. https://www.washingtonpost.com/politics/2021/06/24/most-brutal-debunking-trumps-fraud-claims-yet-republicans/. Accessed 15 Dec 2021
7. Blinken T (2021) The United States officially rejoins the Paris Agreement. Press Statement Antony J. Blinken, Secretary of State; February 19, 2021. https://www.state.gov/the-united-states-officially-rejoins-the-paris-agreement/. Accessed 15 Dec 2021
8. Bourdieu P, Wacquant LJD (1992) An invitation to reflexive sociology. University of Chicago Press, Chicago
9. Brennan Center for Justice (2017) Debunking the election fraud myth. New York University School of Law. https://www.brennancenter.org/sites/default/files/analysis/Briefing_Memo_Debunking_Voter_Fraud_Myth.pdf. Accessed 15 Dec 2021
10. Burchill S (2005) The national interest in international relations theory. Palgrave McMillan, Houndmills
11. Center for Disease Control (2021) Ensuring Covid-19 vaccines work, November 10, 2021. https://www.cdc.gov/coronavirus/2019-ncov/vaccines/effectiveness/index.html. Accessed 15 Dec 2021
12. Council on Foreign Relations (2021) https://www.cfr.org/global-conflict-tracker/conflict/civil-war-libya. Accessed 15 Dec 2021
13. de Jouvenel B (1958) Authority: the efficient imperative. In: Friedrich CJ (ed) Authority. Harvard University Press, Cambridge, Massachusetts
14. Duit A, Galaz V (2008) Governance and complexity—emerging issues for governance theory. Governance 21(3):311–335. https://doi.org/10.1111/j.1468-0491.2008.00402.x. Accessed 5 May 2021
15. Elster J (1989) Nuts and bolts for the social sciences. Cambridge University Press, Cambridge
16. Emirbayer M (1997) Manifesto for a relational sociology. Am J Sociol 103(2):281–317. http://links.jstor.org/sici?sici=0002-9602%28199709%29103%3A2%3C281%3AMFARS%3E2.0.CO%3B2-A. Accessed 30 Sept 2020
17. Fitzgerald FS (2004) The great gatsby. Scribner, New York
18. Foucault M (1990) The history of sexuality. Vol l, An Introduction (trans by Hurley R). Vintage Books, New York
19. Galbraith J (1983) The anatomy of power. Houghton Mifflin Company, Boston
20. Haxton B (2001) Heraclitus: fragments. Penguin Books, New York
21. Johns Hopkins Medicine (2021) What is coronavirus? https://www.hopkinsmedicine.org/health/conditions-and-diseases/coronavirus. Accessed 15 Dec 2021
22. Kaufmann W (1996) [Translator] Martin Buber's I and Thou. Simon and Schuster, New York
23. Mandelbaum M (1955) Societal facts. In: O'Neill J (ed) Modes of methodological individualism and collectivism. Heinemann, London, pp 221–247
24. McGrath M (2020) Climate change: US formally withdraws from Paris agreement. BBC; November 4, 2020. https://www.bbc.com/news/science-environment-54797743. Accessed 15 Dec 2021
25. Mearsheimer J (2014) Why is Ukraine the West's fault? UnCommon core: the causes and consequences of the Ukraine crisis. University of Chicago. https://www.youtube.com/watch?app=desktop&v=JrMiSQAGOS4
26. O'Neill J (1973) Modes of individualism and collectivism. Heinemann, London
27. Partridge P (1963) Some notes on the concept of power. Polit Stud 11(2):107–125. https://doi.org/10.1111/j.1467-9248.1963.tb01054.x. Accessed 23 Sept 2020
28. Preiser R, Biggs R, De Vos A, Folke C (2018) Social-ecological systems as complex adaptive systems: organizing principles for advancing research methods and approaches. Ecol Soc 23(4). Resilience Alliance Inc. https://www.jstor.org/stable/26796889. Accessed 5 May 2021
29. Roth A (2022) Vladimir Putin puts Russia's nuclear deterrence forces on high alert. The Guardian, February 27, 2022. https://www.theguardian.com/world/2022/feb/27/vladimir-putin-puts-russia-nuclear-deterrence-forces-on-high-alert-ukraine
30. The King James Version of the Bible (2021) Bible gateway. https://www.biblegateway.com/passage/?search=Ecclesiastes%201&version=KJV. Accessed 16 Dec 2021

31. Toh M, Ogura J, Humayun H, McGee C, Yee I, Cheung E, Fossum S, Kennedy N (2022) The list of global sanctions on Russia for the war in Ukraine. CNN Business, February 25, 2022. https://www.cnn.com/2022/02/25/business/list-global-sanctions-russia-ukraine-war-intl-hnk/index.html
32. Troianovski A (2021) What's driving Putin's Ukraine brinkmanship? New York Times; December 5, 2021. https://www.nytimes.com/2021/12/05/world/europe/putin-russia-ukraine-troops.html. Accessed 15 Dec 2021
33. Tzu L (1963) Tao Te Ching (trans by Lau D). Penguin Group, London
34. Watkins J (1952) Ideal types and historical explanations. In: O'Neill (ed) Modes of individualism and collectivism 1973. Heinemann, London
35. Watson Institute (2021) Costs of war project. Brown University. https://watson.brown.edu/costsofwar/figures/2021/human-and-budgetary-costs-date-us-war-afghanistan-2001-2022. Accessed 15 Dec 2021
36. Wedekind K (2020) Deterrence and third party amplification: the case of Hezbollah and Al Manar. In: Filippidou A (ed) Deterrence. Advanced sciences and technologies for security applications. Springer, Cham. https://doi.org/10.1007/978-3-030-29367-3_8. Accessed 22 June 2021
37. Wendt A (1992) Anarchy is what States make of it: the social construction of power politics. Int Organ 46(2):391–425. The MIT Press. http://www.jstor.org/stable/2706858. Accessed 3 Aug 2020
38. Wendt A (2017) Social theory of international politics. Cambridge University Press, Cambridge
39. World Health Organization (2021) Covid dashboard. https://covid19.who.int/. Accessed 15 Dec 2021
40. Wrong D (1979) Power: its forms, bases, and uses. Transaction Publishers, London and New Brunswick
41. Zahle J (2016) Methodological holism in the social sciences. In: Zalta EN (ed) The Stanford encyclopaedia of philosophy (Summer 2016 Edition). https://plato.stanford.edu/archives/sum2016/entries/holism-social/. Accessed 28 Apr 2021

The manufacturer's authorised representative in the EU is Springer Nature Customer Service Centre GmbH, Europaplatz 3, 69115 Heidelberg, Germany. If you have any concerns regarding our products, please contact ProductSafety@springernature.com

Printed and bound by CPI Group (UK) Ltd, Croydon, CR0 4YY

30/09/2026

02225604-0013